Renewable Energy

Wind Energy. Electron Jet Generators and Propulsions

Alexander Bolonkin

USA, Lulu, 2017

Wind Energy. Electron Jet Generators and Propulsions (Collection of author Reseaches).
Author: Alexander Bolonkin
ISBN 978-1-365-84732-5

In given book considered the topics: utilization wind energy at high altitude, transwer of energy from airborne wind turbines to ground surface, new non turbine electron wind and water electric generators and propulsion system.

Author offers a new method of getting electric energy from wind. A special injector injects electrons into the atmosphere. Wind picks up the electrons and moves them in the direction of wind which is also against the direction of electric field. At some distance from injector a unique grid acquires the electrons, thus charging and producing electricity. This method does not require, as does other wind energy devices, strong columns, wind turbines, or electric generators. This proposed wind installation is cheap. The area of wind braking may be large and produces a great deal of energy. Although this electron wind installations may be in a city, the population will not see them.

Author offers a new high efficiency propulsion non turbine system using electrons for acceleration of the craft. As this system does not heat the air, it does not have the heating limitations of conventional air ramjet hypersonic engines. Offered engine can produce a thrust from a zero flight speed up to the desired escape velocity for space launch. It can work in any planet atmosphere (gas, liquid) and at high altitude.

Copyright@2017 by author
Publisher: USA, Lulu, www.lulu.com

Content:

Abstract
 Introduction. Wind energy in World.
1. Utilization of Wind Energy at High Altitude.
2. Energy Transver from Airborne Wind Turbine.
3. Electrostatic Generator and Electric transfomer.
4. Jet Generator.
5. Wireless Transfer Electricity from Continente to Continent
6. Electron Wind Generator.
7. Electron Air Hypersonic Propulsion.
8. **Electric Hypersonic Space Aircraft**
9. Electron Hydro Electric Generator.
10. Electron Super Speed Hydro Propulsion.

Abstract

In given book considered the topics: utilization wind energy at high altitude, transwer of energy from airborne wind turbines to ground surface, new non turbine electron wind and water electric generators and propulsion system.

Ground based, wind energy extraction systems have reached their maximum capability. The limitations of current designs are: wind instability, high cost of installations, and small power output of a single unit. The wind energy industry needs of revolutionary ideas to increase the capabilities of wind installations. This book suggests a revolutionary innovation which produces a dramatic increase in power per unit and is independent of prevailing weather and at a lower cost per unit of energy extracted. The main innovation consists of large free-flying air rotors positioned at high altitude for power and air stream stability, and an energy transmission system between the air rotor and a ground.

Author offers a new method of getting electric energy from wind. A special injector injects electrons into the atmosphere. Wind picks up the electrons and moves them in the direction of wind which is also against the direction of electric field. At some distance from injector a unique grid acquires the electrons, thus charging and producing electricity. This method does not require, as does other wind energy devices, strong columns, wind turbines, or electric generators. This proposed wind installation is cheap. The area of wind braking may be large and produces a great deal of energy. Although this electron wind installations may be in a city, the population will not see them.

Author offers a new high efficiency propulsion non turbine system using electrons for acceleration of the craft. As this system does not heat the air, it does not have the heating limitations of conventional air ramjet hypersonic engines. Offered engine can produce a thrust from a zero flight speed up to the desired escape velocity for space launch. It can work in any planet atmosphere (gas, liquid) and at high altitude.

Introduction. Wind Energy in World.

Current wind power is the use of air flow through wind turbines to mechanically power generators for electric power. Wind power, as an alternative to burning fossil fuels, is plentiful, renewable, widely distributed, clean, produces no greenhouse gas emissions during operation, consumes no water, and uses little land. The net effects on the environment are far less problematic than those of nonrenewable power sources.

As of 2015, Denmark generates 40% of its electric power from wind, and at least 83 other countries around the world are using wind power to supply their electric power grids.[16] In 2014 global wind power capacity expanded 16% to 369,553 MW. Yearly wind energy production is also growing rapidly and has reached around 4% of worldwide electric power usage,[18] 11.4% in the EU.

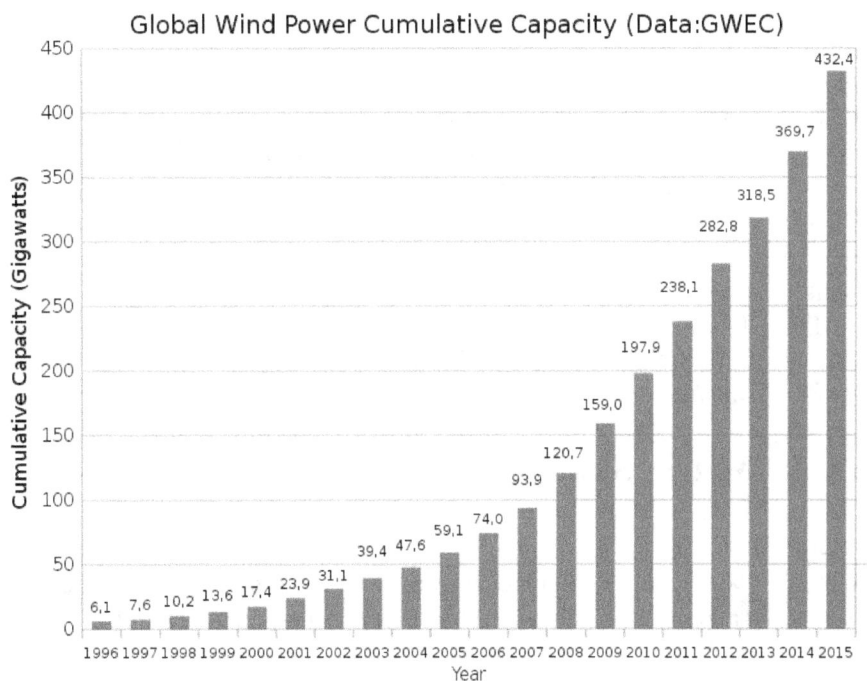

Solar power tends to be complementary to wind.[107][108] On daily to weekly timescales, high pressure areas tend to bring clear skies and low surface winds, whereas low pressure areas tend to be windier and cloudier. On seasonal timescales, solar energy peaks in summer, whereas in many areas wind energy is lower in summer and higher in winter. Thus the seasonal variation of wind and solar power tend to cancel each other somewhat. In 2007 the Institute for Solar Energy Supply Technology of the University of Kassel pilot-tested a combined power plant linking solar, wind, biogas and hydrostorage to provide load-following power around the clock and throughout the year, entirely from renewable sources.[110]

Wind turbines reached grid parity (the point at which the cost of wind power matches traditional sources) in some areas of Europe in the mid-2000s, and in the US around the same time. Falling prices continue to drive the levelized cost down and it has been suggested that it has reached general grid parity in Europe in 2010, and will reach the same point in the US around 2016 due to an expected reduction in capital costs of about 12%.

Chapter 1

Utilization of Wind Energy at High Altitude[*]

Abstract

Ground based, wind energy extraction systems have reached their maximum capability. The limitations of current designs are: wind instability, high cost of installations, and small power output of a single unit. The wind energy industry needs of revolutionary ideas to increase the capabilities of wind installations. This chapter suggests a revolutionary innovation which produces a dramatic increase in power per unit and is independent of prevailing weather and at a lower cost per unit of energy extracted. The main innovation consists of large free-flying air rotors positioned at high altitude for power and air stream stability, and an energy cable transmission system between the air rotor and a ground based electric generator. The air rotor system flies at high altitude up to 14 km. A stability and control is provided and systems enable the changing of altitude.

This chapter includes six examples having a high unit power output (up to 100 MW). The proposed examples provide the following main advantages: 1. Large power production capacity per unit - up to 5,000-10,000 times more than conventional ground-based rotor designs; 2. The rotor operates at high altitude of 1-14 km, where the wind flow is strong and steady; 3. Installation cost per unit energy is low. 4. The installation is environmentally friendly (no propeller noise).

---

[*] Presented as Bolonkin's papers in International Energy Conversion Engineering Conference at Providence., RI, Aug.16-19. 2004. AIAA-2004-5705, AIAA-2004-5756, USA.
Keywords: *wind energy, cable energy transmission, utilization of wind energy at high altitude, air rotor, windmills, Bolonkin.*

Nomenclature (in Metric System)

A - front area of rotor [m^2];
$\alpha = 0.1 - 0.25$ exponent of wind coefficient. One depends from Earth's surface roughness;
A_a - wing area is served by aileron for balance of rotor (propeller) torque moment [m^2];
A_w - area of the support wing [m^2];
C - retail price of 1 kWh [$];
c - production cost of 1 kWh [$];
C_L - lift coefficient (maximum $C_L \approx 2.5$);
C_D – drag coefficient;
$\Delta C_{L,a}$ - difference of lift coefficient between left and right ailerons;
D – drag force [N];
D_r - drag of rotor [N];
E - annual energy produced by flow installation [J];
F – annual profit [$];
$H_o = 10$ m - standard altitude of ground wind installation [m];
H - altitude [m];
I - cost of Installation [$];
K_1 - life time (years);
K_2 – rotor lift coefficient (5-12 [kg/kW]);
L - length of cable [m];
L_y – lift force of wing [N];
M – annual maintenance [$];
N– power [W, joule/sec];

N_o - power at H_o;
r - distance from center of wing to center of aileron [m];
R - radius of rotor (turbine)[m];
S - cross-section area of energy transmission cable [m²];
V - annual average wind speed [m/s];
V_o - wind speed at standard altitude 10 m [m/s](V_o= 6 m/s);
W - weight of installation (rotor + cables)[kg];
W_y – weight of cable [kg];
γ - specific density of cable [kg/m³];
η - efficiency coefficient;
θ - angle between main (transmission) cable and horizontal surface;
λ - ratio of blade tip speed to wind speed;
v - speed of transmission cable [m/s];
ρ - density of flow, ρ=1.225 kg/m³ for air at sea level altitude H = 0; ρ=0.736 at altitude H =5 km;
ρ= 0.413 at H =10 km;
σ - tensile stress of cable [N/m²].

Introduction

Wind is a clean and inexhaustible source of energy that has been used for many centuries to grind grain, pump water, propel sailing ships, and perform other work. Wind farm is the term used for a large number of wind machines clustered at a site with persistent favorable winds, generally near mountain passes. Wind farms have been erected in New Hampshire, in the Tehachapi Mountains. at Altamont Pass in California, at various sites in Hawaii, and may other locations. Machine capacities range from 10 to 500 kilowatts. In 1984 the total energy output of all wind farms in the United States exceeded 150 million kilowatt-hours.

A program of the United States Department of Energy encouraged the development of new machines, the construction of wind farms, and an evaluation of the economic effect of large-scale use of wind power.

The utilization of renewable energy ('green' energy) is currently on the increase. For example, a lot of wind turbines are being installed along the British coast. In addition, the British government has plans to develop off-shore wind farms along their coast in an attempt to increase the use of renewable energy sources. A total of $2.4 billion was injected into renewable energy projects over the last three years in an attempt to meet the government's target of using renewable energy to generate 10% of the country's energy needs by 2010.

This British program saves the emission of almost a millions tons of carbon dioxide. Denmark plans to get about 30% of their energy from wind sources.

Unfortunately, current wind energy systems have deficiencies which limit their commercial applications:

1. Wind energy is unevenly distributed and has relatively low energy density. Huge turbines cannot be placed on the ground, many small turbines must be used instead. In California, there are thousands of small wind turbines. However, while small turbines are relatively inefficient, very huge turbines placed at ground are also inefficient due to the relatively low wind energy

density and their high cost. The current cost of wind energy is higher then energy of thermal power stations.
2. Wind power is a function of the cube of wind velocity. At surface level, wind has low speed and it is non-steady. If wind velocity decreases in half, the wind power decreases by a factor of 8 times.
3. The productivity of a wind-power system depends heavily on the prevailing weather.
4. Wind turbines produce noise and visually detract from the landscape.

Figure 1. Ground wind engines

There are many research programs and proposals for the wind driven power generation systems, however, all of them are ground or tower based. System proposed in this article is located at high altitude (up to the stratosphere), where strong permanent and steady streams are located. The also proposes a solution to the main technologist challenge of this system; the transfer of energy to the ground via a mechanical transmission made from closed loop, modern composite fiber cable.

The reader can find the information about this idea in [1], the wind energy in references [2]-[3], a detailed description of the innovation in [4]-[5], and new material used in the proposed innovation in [6]-[9]. The application of this innovation and energy transfer concept to other fields can be found in [10]-[19].

Description of Innovation

Main proposed high altitude wind system is presented in
Figure 1. That includes: rotor (turbine) 1, support wing 2, cable mechanical transmission and keep system 3, electro-generator 4, and stabilizer 5. The transmission system has three cables (Figure1e): main (central) cable, which keeps the rotor at a given altitude, and two transmission mobile cables, which transfer energy from the rotor to the ground electric generator. The device of Figure1f allows changing a cable length and a rotor altitude. In calm weather the rotor can be support at altitude by dirigible 9 (Figure1c) or that is turned in vertical position and support by rotation from the electric generator (Figure1d). If the wind is less of a minimum speed for support of rotor at altitude the rotor may be supported by autogiro mode in position of Figure1d. The probability of full wind calm at a high altitude is small and depends from an installation location.

Figure 2 shows other design of the proposed high altitude wind installation. This rotor has blades, 10, connected to closed-loop cables. The forward blades have a positive angle and lift force. When they are in a back position the lift force equals zero. The rotor is supported at the high altitude by the blades and the wing 2 and stabilizer 5. That design also has energy transmission 3 connected to the ground electric generator 4.

Figure 3 shows a parachute wind high altitude installation. Here the blades are changed by parachutes. The parachutes have a large air drag and rotate the cable rotor 1. The wind 2 supports the installation in high altitude. The cable transmission 3 passes the rotor rotation to the ground electric generator 4.

A system of Figure 4 uses a large Darries air turbine located at high altitude. This turbine has four blades. The other components are same with previous projects.

Wind turbine of Figure 5 is a wind ground installation. Its peculiarity is a gigantic cable-blade rotor. That has a large power for low ground wind speed. It has four columns with rollers and closed-loop cable rotor with blades 10. The wind moves the blades, the blades move the cable, and the cable rotates an electric generator 4.

Problems of Launch, Start, Guidance, Control, Stability, and Others

Launching. It is not difficult to launch the installations having support wing or blades as described in Figure 1-4. If the wind speed is more than the minimum required speed (>2-3 m/s), the support wing lifts the installation to the desired altitude.

Starting. All low-speed rotors are self-starting. All high-speed rotors (include the ground rotor of Figure 5) require an initial starting rotation from the ground motor-generator 4 (figures 1,5).

Guidance and Control. The control of power, revolutions per minute, and torque moment are operated by the turning of blades around the blade longitudinal axis. The control of altitude may be manual or automatic when the wind speed is normal and over admissible minimum. Control is effected by wing flaps and stabilizer (elevator), fin, and ailerons (figs. 1,2,4).

Stability. Stability of altitude is produced by the length of the cable. Stability around the blade longitudinal axis is made by stabilizer (see figs.1,2,4). Rotor directional stability in line with the flow can be provided by fins (figs. 1). When the installation has the support wing rigidly connected to the rotor, the stability is also attained by the correct location of the center of gravity of the installation (system rotor-wing) and the point of connection of the main cable and the tension elements. The center-of-gravity and connection point must be located within a relatively narrow range 0.2-0.4 of the average aerodynamic chord of the support wing (for example, see Figure 1). There is the same requirement for the additional support wings such as Figure 2-4.

Torque moment is balanced by transmission and wing ailerons (see figs.1-4).

The wing lift force, stress of main cable are all regulated automatic by the wing flap or blade stabilizer.

The location of the installation of Figure 2 at a given point in the atmosphere may be provided by tension elements shown on Figure 2. These tension elements provide a turning capability for the installation of approximately $\pm 45^0$ degrees in the direction of flow (see. Figure 2.).

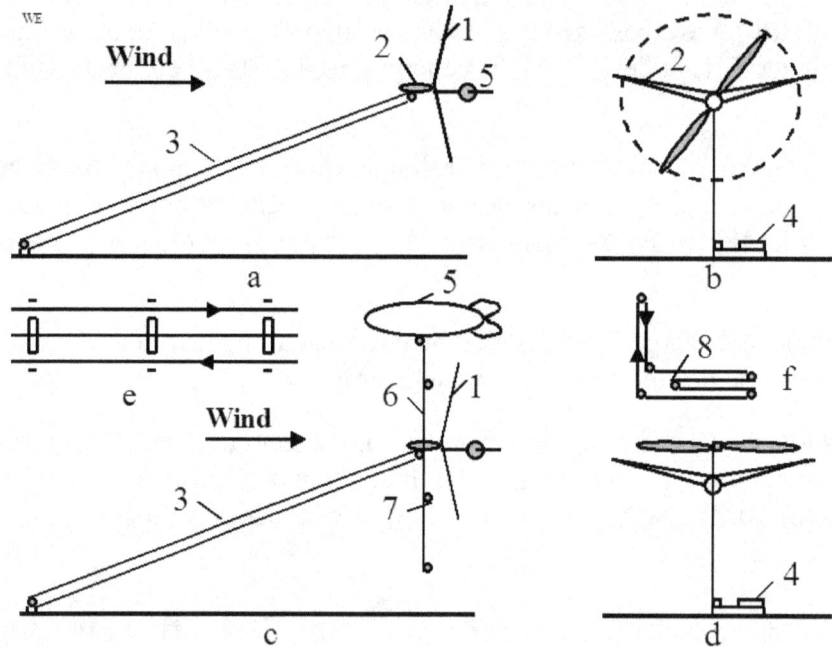

Figure 1a. Propeller high altitude wind energy installation and cable energy transport system. Notation: a – side view; 1 – wind rotor; 2 – wing with ailerons; 3 – cable energy transport system; 4 – electric generator; 5 – stabilizer; b – front view; c – side view with a support dirigible 9, vertical cable 6, and wind speed sensors 7; d - keeping of the installation at a high altitude by rotate propeller; e – three lines of the transmission - keeper system. That includes: main (central) cable and two mobile transmission cables; f – energy transport system with variable altitude; 8 – mobile roller.

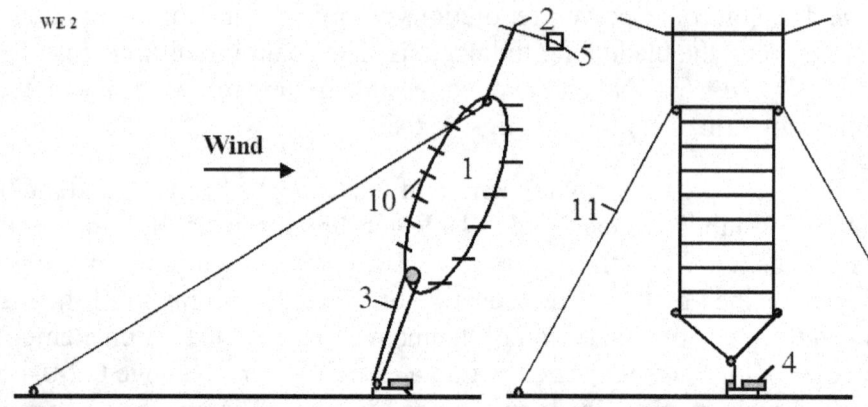

Figure 2. High altitude wind energy installation with the cable turbine. Notation: 10 – blades; 11 – tensile elements (bracing)(option).

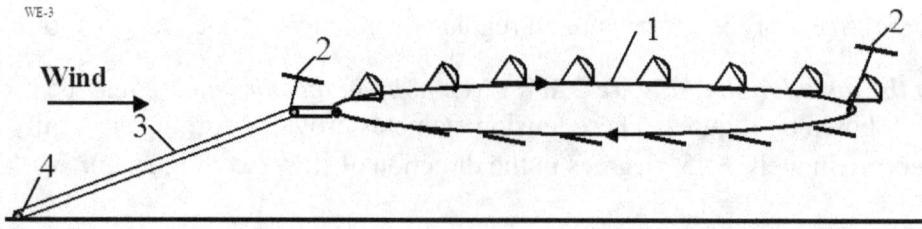

Figure 3. High altitude wind energy installation with the parachute turbine.

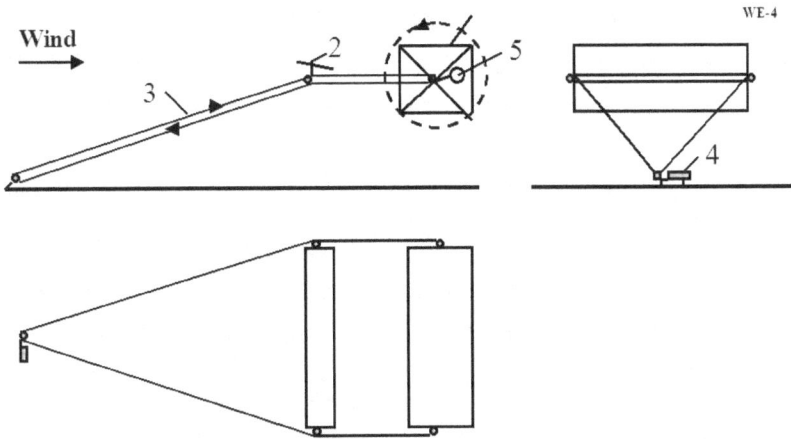

Figure 4 (right). High altitude wind energy installation with Darrieus turbine.

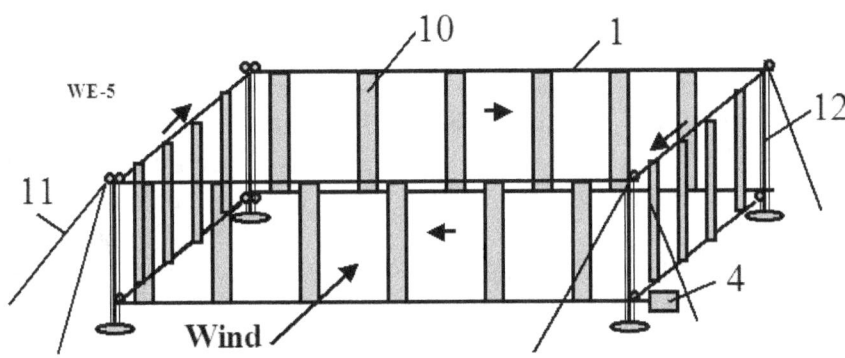

Figure 5. Ground wind cable rotor of a large power.

Minimum wind speed. The required minimum wind-speed for most of the suggested installation designs is about 2 m/s. The probability of this low wing speed at high altitude is very small (less 0.001). This minimum may be decreased still further by using the turning propeller in an autogiro mode. If the wind speed is approximately zero, the rotor can be supported in the atmosphere by a balloon (dirigible) as is shown on Figure1c or a propeller rotated by the ground power station as is shown on Figure1d. The rotor system may also land on the ground and start again when the wind speed attains the minimum speed for flight.

A Gusty winds. Large pulsations of wind (aerodynamic energy) can be smoothed out by inertial flywheels.

The suggested Method and Installations for utilization of wind energy has following peculiarities from current conventional methods and installations:

1. Proposed installation allows the collection of energy from a large area – tens and even hundreds of times more than conventional wind turbines. This is possible because an expensive tower is not needed to fix our rotor in space. Our installation allows the use of a rotor with a very large diameter, for example 100-200 meters or more.

2. The proposed wind installations can be located at high altitude 100 m - 14 km. The wind speeds are 2-4 times faster and more stable at high altitude compared to ground surface winds used by the altitude of conventional windmills (10-70 meters of height). In certain geographic areas high altitude wind flows have a continuous or permanent nature. Since wind power increases at the cube of wind speed, wind rotor power increases by 27 times when wind speed increases by 3 times.
3. In proposed wind installation the electric generator is located at ground. There are proposals where electric generator located near a wind rotor and sends electric current to a ground by electric wares. However, our rotor and power are very large (see projects below). Proposed installations produce more power by thousands of times compared to the typical current wind ground installation (see point 1, 2 above). The electric generator of 20 MW weighs about 100 tons (specific weigh of the conventional electric generator is about 3-10 kg/kW). It is impossible to keep this weigh by wing at high altitude for wind speed lesser then 150 m/s.
4. One of the main innovations of the given invention is the *cable transfer* (transmission) of energy from the wind rotor located at high altitude to the electric generator located on ground. In proposed Installation it is used a new cable transmission made from artificial fibers. This transmission has less a weigh in thousands times then copper electric wires of equal power. The wire having diameter more 5 mm passes 1-2 ampere/sq.mm. If the electric generator produces 20 MW with voltage 1000 Volts, the wire cross-section area must be 20,000 mm^2, (wire diameter is160 mm). The cross-section area of the cable transmission of equal power is only 37 mm^2 (cable diameter 6.8 mm^2 for cable speed 300 m/s and admissible stress 200 kg/mm^2, see Project 1). The specific weight of copper is 8930 kg/m^3, the specific weight of artificial fibers is 1800 kg/m^3. If the cable length for altitude 10 km is 25 km the double copper wire weighs 8930 tons (!!), the fiber transmission cable weighs only 3.33 tons. It means the offered cable transferor energy of equal length is easier in 2682 times, then copper wire. The copper wires is very expensive, the artificial fiber is cheap.

All previous attempts to place the generator near the rotor and connect it to ground by electric transmission wires were not successful because the generator and wires are heavy.

Some Information about Wind Energy

The power of a wind engine strongly depends on the wind speed (to the third power). Low altitude wind ($H = 10$ m) has the standard average speed $V = 6$ m/s. High altitude wind is powerful and that has another important advantage, it is stable and constant. This is true practically everywhere.

Wind in the troposphere and stratosphere are powerful and permanent. For example, at an altitude of 5 km, the average wind speed is about 20 M/s, at an altitude 10-12 km the wind may reach 40 m/s (at latitude of about 20-35^{0}N).

There are permanent jet streams at high altitude. For example, at H = 12-13 km and about 25^{0}N latitude. The average wind speed at its core is about 148 km/h (41 m/s). The most intensive portion, with a maximum speed 185 km/h (51 m/s) latitude 22^0, and 151 km/h (42 m/s) at latitude 35^0 in North America. On a given winter day, speeds in the jet core may exceed 370 km/h (103 m/s) for a distance of several hundred miles along the direction of the wind. Lateral wind shears in the direction normal to the jet stream may be 185 km/h per 556 km to right and 185 km/h per 185 km to the left.

The wind speed of $V = 40$ m/s at an altitude $H = 13$ km provides 64 times more energy than surface wind speeds of 6 m/s at an altitude of 10 m.

This is a gigantic renewable and free energy source. (See reference: *Science and Technolody,v.2, p.265)*.

Cable Transmission Energy Problem

The primary innovations presented in this paper are locating the rotor at high altitude, and an energy transfer system using a cable to transfer mechanical energy from the rotor to a ground power station. The critical factor for this transfer system is the weight of the cable, and its air drag.

Twenty years ago, the mass and air drag of the required cable would not allow this proposal to be possible. However, artificial fibers are currently being manufactured, which have tensile strengths of 3-5 times more than steel and densities 4-5 times less then steel. There are also experimental fibers (whiskers) which have tensile strengths 30-100 times more than a steel and densities 2 to 5 times less than steel. For example, in the book [6] p.158 (1989), there is a fiber (whisker) C_D, which has a tensile strength of σ = 8000 kg/mm² and density (specific gravity) of γ = 3.5 g/cm³. If we use an estimated strength of 3500 kg/mm² (σ =7·10¹⁰ N/m², γ = 3500 kg/m³), then the ratio is γ/σ = 0.1×10⁻⁶ or σ/γ = 10×10⁶. Although the described (1989) graphite fibers are strong (σ/γ = 10×10⁶), they are at least still ten times weaker than theory predicts. A steel fiber has a tensile strength of 5000 MPA (500 kg/sq.mm), the theoretical limit is 22,000 MPA (2200 kg/mm²)(1987); the polyethylene fiber has a tensile strength 20,000 MPA with a theoretical limit of 35,000 MPA (1987). The very high tensile strength is due to its nanotubes structure.

Apart from unique electronic properties, the mechanical behavior of nanotubes also has provided interest because nanotubes are seen as the ultimate carbon fiber, which can be used as reinforcements in advanced composite technology. Early theoretical work and recent experiments on individual nanotubes (mostly MWNT's, Multi Wall Nano Tubes) have confirmed that nanotubes are one of the stiffest materials ever made. Whereas carbon-carbon covalent bonds are one of the strongest in nature, a structure based on a perfect arrangement of these bonds oriented along the axis of nanotubes would produce an exceedingly strong material. Traditional carbon fibers show high strength and stiffness, but fall far short of the theoretical, in-plane strength of graphite layers by an order of magnitude. Nanotubes come close to being the best fiber that can be made from graphite.

For example, whiskers of Carbon nanotube (CNT) material have a tensile strength of 200 Giga-Pascals and a Young's modulus over 1 Tera Pascals (1999). The theory predicts 1 Tera Pascals and a Young's modules of 1-5 Tera Pascals. The hollow structure of nanotubes makes them very light (the specific density varies from 0.8 g/cc for SWNT's (Single Wall Nano Tubes) up to 1.8 g/cc for MWNT's, compared to 2.26 g/cc for graphite or 7.8 g/cc for steel).

Specific strength (strength/density) is important in the design of the systems presented in this paper; nanotubes have values at least 2 orders of magnitude greater than steel. Traditional carbon fibers have a specific strength 40 times that of steel. Since nanotubes are made of graphitic carbon, they have good resistance to chemical attack and have high thermal stability. Oxidation studies have shown that the onset of oxidation shifts by about 100⁰ C or higher in nanotubes compared to high modulus graphite fibers. In a vacuum, or reducing atmosphere, nanotube structures will be stable to any practical service temperature.

The artificial fibers are cheap and widely used in tires and everywhere. The price of SiC whiskers produced by Carborundum Co. with σ=20,690 MPa and γ=3.22 g/cc was $440 /kg in 1989. The market price of nanotubes is too high presently (~$200 per gram)(2000). In the last 2-3 years, there have been several companies that were organized in the US to produce and market nanotubes. It is anticipated that in the next few years, nanotubes will be available to consumers for less than $100/pound.

Table 1. Material properties

Material Whiskers	Tensile strength kg/mm²	Density g/cm³	Fibers	Tensile strength kg/mm²	Density g/cm³
AlB_{12}	2650	2.6	QC-8805	620	1.95
B	2500	2.3	TM9	600	1.79
B_4C	2800	2.5	Thorael	565	1.81
TiB_2	3370	4.5	Allien 1	580	1.56
SiC	1380-4140	3.22	Allien2	300	0.97

Reference [6]-[9].

Below, the author provides a brief overview of recent research information regarding the proposed experimental (tested) fibers. In addition, the author also addresses additional examples, which appear in these projects and which can appear as difficult as the proposed technology itself. The author is prepared to discuss the problems with organizations which are interested in research and development related projects.

Industrial fibers with σ = 500-600 kg/mm², γ = -1800 kg/m³, and σ/γ = 2,78x10⁶ are used in all our projects (safety σ =200-250 kg/mm²)(see below).

Brief Theory of Estimation of Suggested Installations

Rotor

Power of a wind energy N [Watt, Joule/sec]

$$N = 0.5 \eta \rho A V^3 \quad [W] \tag{1}$$

The coefficient of efficiency, η, equals 0.15-0.35 for low speed rotors (ratio of blade tip speed to wind speed equals $\lambda \approx 1$); η = 0.35-0.5 for high speed rotors (λ = 5-7). The Darrieus rotor has η = 0.35 - 0.4. The propeller rotor has η = 0.45-0.50. The theoretical maximum equals η = 0.67.

The energy is produced in one year is (1 year $\approx$ 30.2×10⁶ work sec) [J]

$$E = 3600 \times 24 \times 350 \approx 30 \times 10^6 N \quad [J]. \tag{1'}$$

Wind speed increases with altitude as follows

$$V = (H/H_o)^\alpha V_o, \tag{2}$$

where α = 0.1 - 0.25 exponent coefficient depends from surface roughness. When the surface is water, α = 0.1; when surface is shrubs and woodlands α = 0.25.

Power increases with altitude as the cube of wind speed

$$N = (H/H_o)^{3\alpha} N_o, \tag{3}$$

where N_o is power at H_o.

The drag of the rotor equals

$$D_r = N/V \quad [\text{N}] \tag{4}$$

The lift force of the wing, L_y, is

$$L_y = 0.5 C_L \rho V^2 A_w, \quad [\text{N}], \qquad L_y \approx gW, \tag{5}$$

where C_L is lift coefficient (maximum $C_L \approx 2.5$), A_w is area of the wing, W is weight of installation + 0.5 weight of all cables [kg], $g = 9.81$ m/s^2.

The drag of the wing is

$$D = 0.5 C_D \rho V^2 A_w, \quad [\text{N}] \tag{6}$$

where C_D is the drag coefficient (maximum $C_D \approx 1.2$).

The optimal speed of the parachute rotor equals $1/3 V$ and the theoretical maximum of efficiency coefficient is 0.5.

The annual energy produced by the wind energy extraction installation equals

$$E = 8.33\, N \quad [\text{kWh}] \tag{7}$$

Cable Energy Transfer, Wing Area, and other Parameters

Cross-section area of transmission cable, S, is

$$S = N/v\sigma, \quad [\text{m}^2], \tag{8}$$

Cross-section area of main cable, S_m, is

$$S_m = (D_r + D)/\sigma, \quad [\text{m}^2], \tag{8'}$$

Weight of cable is

$$W_r = SL\gamma, \quad [\text{kg}], \tag{9}$$

The production cost, c, in kWh is

$$c = \frac{M + I/K_1}{E}, \tag{10}$$

The annual profit

$$F = (C-c)E. \tag{11}$$

The required area [m^2] of the support wing is

$$A_w = \frac{\eta A \sin\theta}{C_L} \tag{12}$$

where θ is the angle between the support cable and horizontal surface.

The wing area is served by ailerons for balancing of the rotor (propeller) torque moment

$$A_a = \frac{\eta A R}{\lambda_i \Delta C_{L,a} r} \tag{13}$$

The minimum wind speed [m/s] for installation support by the wing alone is

$$V_{min} = \sqrt{\frac{2W}{C_{L,max} \rho A_w}} \tag{14}$$

where W is the total weight of the airborne system including transmission. If a propeller rotor is used in a gyroplane mode, minimal speed will decrease by 2-2,5 times. If wind speed equals zero, the required power for driving the propeller in a propulsion (helicopter) mode is

$$N_s = W/K_2 \quad [\text{kW}], \tag{15}$$

The specific weight of energy storage (flywheel) can be estimated by

$$E_s = \sigma/2\gamma \quad [\text{J/kg}]. \tag{16}$$

For example, if $\sigma = 200$ kg/mm², $\gamma = 1800$ kg/m³, then $E_s = 0.56$ MJ/kg or $E_s = 0.15$ kWh/kg.

For comparison of the different ground wind installations their efficiency and parameters are computed for the standard wind conditions: the wind speed equals $V = 6$ m/s at the altitude $H = 10$ m.

Projects

Project 1

High-Speed Air Propeller Rotor (Figure 1)
For example, let us consider a rotor diameter of 100 m ($A = 7850$ m²), at an altitude $H = 10$ km ($\rho = 0.4135$ kg/m³), wind speed of $V = 30$ m/s, an efficiency coefficient of $\eta = 0.5$, and a cable tensile stress of $\sigma = 200$ kg/mm².

Then the power produced is $N = 22$ MW [Eq. (1)], which is sufficient for city with a population of 250,000. The rotor drag is $D_r = 73$ tons [Eq.(4)], the cross-section of the main cable area is $S = 1.4 D_r/\sigma = 1.35 \times 73/0.2 \approx 500$ mm², the cable diameter equals $d = 25$ mm; and the cable weight is $W = 22.5$ tons [Eq.(9)] (for $L = 25$ km). The cross-section of the transmission cable is $S = 36.5$ mm² [Eq.(8)], $d = 6.8$ mm, weight of two transmission cables is $W = 3.33$ tons for cable speed $v = 300$ m/s [Eq.(9)].

The required wing size is 20×100 m ($C_L = 0.8$) [Eq.(12)], wing area served by ailerons is 820 sq.m [Eq.(13)]. If $C_L = 2$, the minimum speed is 2 m/s [Eq.(14)].

The installation will produce an annual energy $E = 190$ GWh [Eq.(7)]. If the installation cost is $200K, has a useful life of 10 years, and requires maintenance of $50K per year, the production cost is $c = 0.37$ cent per kWh [Eq(10)]. If retail price is $0.15 per kWh, profit $0.1 per kWh, the total annual profit is $19 millions per year [Eq.(11)].

The Project #2

Large Air Propeller at Altitude H = 1 km (Figure 1)

Let us consider a propeller diameter of 300 m, with an area $A = 7 \times 10^4$ m^2, at an altitude $H = 1$ km, and a wind speed of 13 m/s. The average blade tip speed is 78 m/s.

The full potential power of the wind streamer flow is 94.2 MW. If the coefficient of efficiency is 0.5 the useful power is $N = 47.1$ MW. For other wind speed. the useful power is: $V = 5$ m/s, $N = 23.3$ MW; $V = 6$ m/s, $N = 47.1$ MW; $V = 7$ m/s, $N = 74.9$ MW; $V = 8$ m/s, $N = 111.6$ MW; $V = 9$ m/s, $N = 159$ MW; $V = 10$ m/s, $N = 218$ MW.

Estimation of Economical Efficiency

Let us assume that the cost of the Installation is $3 million, a useful life of 10 years, and request maintenance of $100,000/year. The energy produced in one year is $E = 407$ GWh [Eq.(7)]. The basic cost of energy is $0.01 /kWh.

The Some Technical Parameters

Altitude $H = 1$ km
The drag is about 360 tons. Ground connection (main) cable has cross-section area of 1800 sq,mm [Eq.(8')], $d = 48$ mm, and has a weight of 6480 kg. The need wing area is 60x300 m. The aileron area requested for turbine balance is 6740 sq.m.

If the transmission cable speed is 300 m/s, the cross-section area of transmission cable is 76 sq.mm and the cable weight is 684 kg (composite fiber).

Altitude $H = 13$ km
At an altitude of $H = 13$ km. the air density is $\rho = 0.2666$, and the wind speed is $V = 40$ m/s. The power for efficiency coefficient 0.5 is 301.4 MgW. The drag of the propeller is approximately 754 tons. The connection cable has a cross-sectional area of 3770 sq.mm, a diameter is $d = 70$ mm and a weight of 176 tons. The transmission cable has a sectional area 5 sq.c and a weight of 60 tons (vertical transmission only 12 tons).

The installation will produce energy $E = 2604$ GWh per year. If the installation costs $5 million, maintenance is $200,000/year, and the cost of 1 kWh will be $0.0097/kWh.

Project #3

Air Low Speed Wind Engine with Free Flying Cable Flexible Rotor (Figure 2)

Let us consider the size of cable rotor of width 50 m, a rotor diameter of 1000 m, then the rotor area is $A = 50 \times 1000 = 50,000$ sq.m. The angle rope to a horizon is 70°. The angle of ratio lift/drag is about 2.5°.

The average conventional wind speed at an altitude $H = 10$ m is $V = 6$ m/s. It means that the speed at the altitude 1000 m is 11.4 - 15 m/s. Let us take average wind speed $V = 13$ m/s at an altitude $H = 1$ km.

The power of flow is

$N = 0.5 \rho V^3 A \cos 20^0 = 0.5 \times 1.225 \times 13^3 \times 1000 \times 50 \times 0.94 = 63$ MW.

If the coefficient efficiency is $\eta = 0.2$ the power of installation is

$\eta = 0.2 \times 63 = 12.5$ MW.

The energy 12.5 MW is enough for a city with a population at 150,000.

If we decrease our Installation to a 100x2000 m the power decreases approximately by 6 times (because the area decreases by 4 times, wind speed reaches more 15 m/s at this altitude. Power will be 75 MW. This is enough for a city with a population about 1 million of people.

If the average wind speed is different for given location the power for the basis installation will be: $V = 5$ m/s, $N = 7.25$ MW; $V = 6$ m/s, $N = 12.5$ MW; $V = 7$ m/s, $N = 19.9$ MW; $V = 8$ m/s, $N = 29,6$ MW; $V = 9$ m/s, $N = 42.2$ MW; $V = 10$ m/s, $N = 57.9$ MW.

Economical Efficiency

Let us assume that the cost of our installation is $1 million. According to the book "Wind Power" by P. Gipe [2], the conventional wind installation with the rotor diameter 7 m costs $20,000 and for average wind speeds of 6 m/s has power 2.28 kW, producing 20,000 kWh per year. To produce the same amount of power as our installation using by conventional methods, we would need 5482 (12500/2.28) conventional rotors, costing $110 million. Let us assume that our installation has a useful life of 10 years and a maintenance cost is $50,000/year. Our installation produces 109,500,000 kWh energy per year. Production costs of energy will be approximately 150,000/109,500,000 = 0.14 cent/kWh. The retail price of 1 kWh of energy in New York City is $0.15 now. The revenue is 16 millions. If profit from 1 kWh is $0.1, the total profit is more 10 millions per year.

Estimation Some Technical Parameters

The cross-section of main cable for an admissible fiber tensile strange $\sigma = 200$ kg/sq.mm is $S = 2000/0.2 = 10,000$ mm². That is two cable of diameter $d = 80$ mm. The weight of the cable for density 1800 kg/m³ is

$W = SL\gamma = 0.01 \times 2000 \times 1800 = 36$ tons.

Let us assume that the weight of 1 sq.m of blade is 0.2 kg/m² and the weight of 1 m of bulk is 2 kg. The weight of the 1 blade will be 0.2 x 500 = 100 kg, and 200 blades are 20 tons. If the weight of one bulk is 0.1 ton, the weight of 200 bulks is 20 tons.

The total weight of main parts of the installation will be 94 tons. We assume 100 tons for purposes of our calculations.

The minimum wind speed when the flying rotor can supported in the air is (for $C_y = 2$)

$V = (2W/C_y \rho S)^{0.5} = (2 \times 100 \times 10^4 / 2 \times 1.225 \times 200 \times 500)^{0.5} = 2.86$ m/s

The probability of the wind speed falling below 3 m/s when the average speed is 12 m/s, is zero, and for 10 m/s is 0.0003. This equals 2.5 hours in one year, or less than one time per year. The wind at high altitude has greater speed and stability than near ground surface. There is a strong wind at high altitude even when wind near the ground is absent. This can be seen when the clouds move in a sky on a calm day.

Project #4

Low Speed Air Drag Rotor (Figure 3)
Let us consider a parachute with a diameter of 100 m, length of rope 1500 m, distance between the parachutes 300 m, number of parachute 3000/300 = 10, number of worked parachute 5, the area of one parachute is 7850 sq.m, the total work area is $A = 5 \times 7850 = 3925$ sq.m. The full power of the flow is 5.3 MW for $V=6$ m/s. If coefficient of efficiency is 0.2 the useful power is $N = 1$ MW. For other wind speed the useful power is: $V=5$ m/s, $N=0.58$ MW; $V=6$ m/s, $N=1$ MW; $V=7$ m/s, $N=1.59$ MW; $V=8$ m/s, $N=2.37$ MW; $V=9$ m/s, $N=3.375$ MW; $V=10$ m/s, $N=4.63$ MW.

Estimation of Economical Efficiency
Let us take the cost of the installation $0.5 million, a useful life of 10 years and maintenance of $20,000/year. The energy produced in one year (when the wind has standard speed 6 m/s) is $E = 1000 \times 24 \times 360 = 8.64$ million kWh. The basic cost of energy is 70,000/8640,000 = 0.81 cent/kWh.

The Some Technical Parameters
If the thrust is 23 tons, the tensile stress is 200 kg/sq.mm (composed fiber), then the parachute cable diameter is 12 mm, The full weight of the installation is 4.5 tons. The support wing has size 25x4 m.

Project #5

High Speed Air Darreus Rotor at an Altitude 1 km (Figure 4)
Let us consider a rotor having the diameter of 100 m, a length of 200 m (work area is 20,000 m²). When the wind speed at an altitude $H = 10$ m is $V=6$ m/s, then at an altitude $H = 1000$ m it is 13 m/s. The full wind power is 13,46 MW. Let us take the efficiency coefficient 0.35, then the power of the Installation will be $N = 4.7$ MW. The change of power from wind speed is: $V = 5$ m/s, $N = 2.73$ MW; $V = 6$ m/s, $N = 4.7$ MW; $V = 7$ m/s, $N = 7.5$ MW; $V = 8$ m/s, $N = 11.4$ MW; $V = 9$ m/s, $N = 15.9$ MW; $V = 10$ m/s, $N = 21.8$ MW.

At an altitude of $H = 13$ km with an air density 0.267 and wind speed $V = 40$ m/s, the given installation will produce power $N = 300$ MW.

Estimation of Economical Efficiency
Let us take the cost of the Installation at $1 million, a useful life of 10 years, and maintenance of $50,000 /year. Our installation will produce $E = 41$ millions kWh per year (when the wind speed equals 6 m/s at an altitude 10 m). The prime cost will be 150,000/41,000,000 = 0.37 cent/kWh. If the customer price is $0.15/kWh and profit from 1 kWh is $0.10 /kWh the profit will be $4.1 million per year.

Estimation of Technical Parameters
The blade speed is 78 m/s. Numbers of blade is 4. Number of revolution is 0.25 revolutions per second. The size of blade is 200x0.67 m. The weight of 1 blade is 1.34 tons. The total weight of the Installation is about 8 tons. The internal wing has size 200x2.3 m. The additional wing has size 200x14.5 m and weight 870 kg. The cross-section area of the cable transmission having an altitude of $H = 1$ km is 300 sq.mm, the weight is 1350 kg.

Project #6

Ground Wind High Speed Engine (Figure 5)
Let us consider the ground wind installation (Figure5) with size 500x500x50 meters. The work area is 500x50x2 = 50,000 sq.m. The tower is 60 meter tall, the flexible rotor located from 10 m to 60 m. If the wind speed at altitude 10 m is 6 m/s, that equals 7.3 m/s at altitude 40 m.

The theoretical power is

$N_t = 0.5\rho V^3 A = 0.5 \times 1.225 \times 7.3^3 \times 5 \times 10^4 = 11.9$ MgW.

For coefficient of the efficiency equals 0.45 the useful power is

$N = 0.45 \times 11.9 = 5.36$ MW.

For other wind speed at an altitude 6 m/s the useful power is: $V = 5$ m/s, $N = 3.1$ MW; $V = 6$ m/s, $N = 5.36$ MW; $V = 7$ m/s, $N = 8.52$ MW; $V = 8$ m/s, $N = 12.7$ MW; $V = 9$ m/s, $N = 18.1$ MW; $V = 10$ m/s, $N = 24.8$ MW.

Figure 6. Kit wind engine

Economic Estimation
In this installation the rotor will be less expensive than previous installations because the high-speed rotor has a smaller number of blades and smaller blades (see technical data below). However this installation needs 4 high (60 m) columns. Take the cost of the installation at $1 million with a useful life of 10 years. The maintenance is projected at about $50,000 /year.

This installation will produce $E = 5360$ kW x 8760 hours = 46.95 MWh energy (for the annual average wind-speed $V = 6$ m/s at $H = 10$ m). The cost of 1 kWh is 150,000/46,950,000 = 0.4 cent/kWh. If the retail price is $0.15/kWh and delivery cost 30%, the profit is $0.10 per kWh, or $4.7 million per year.

Estimation of Some Technical Parameters
The blade speed is 6 x 7.3 = 44 m/s. The distance between blades is 44 m. The number of blade is 4000/44 = 92.

Discussion and Conclusion

Conventional windmills are approached their maximum energy extraction potential relative to their installation cost. No relatively progress has been made in windmill technology in the last 50 years. The wind energy is free, but its production more expensive then its production in heat electric stations.

Current wind installations cannot essential decrease a cost of kWh, stability of energy production. They cannot increase of power of single energy unit. The renewable energy industry needs revolutionary ideas that improve performance parameters (installation cost and power per unit) and that significantly decreases (in 10-20 times) the cost of energy production. This paper offers ideas that can move the wind energy industry from stagnation to revolutionary potential.

The following is a list of benefits provided by the proposed system compared to current installations:

1. The produced energy at least in 10 times cheaper then energy received of all conventional electric stations includes current wind installation.
2. The proposed system is relatively inexpensive (no expensive tower), it can be made with a very large thus capturing wind energy from an enormous area (hundreds of times more than typical wind turbines).
3. The power per unit of proposed system in some hundreds times more of typical current wind installations.
4. The proposed installation not requires large ground space.
5. The installation may be located near customers and not require expensive high voltage equipment. It is not necessary to have long, expensive, high-voltage transmission lines and substations. Ocean going vessels can use this installation for its primary propulsion source.
6. No noise and bad views.
7. The energy production is more stability because the wind is steadier at high altitude. The wind may be zero near the surface but it is typically strong and steady at higher altitudes. This can be observed when it is calm on the ground, but clouds are moving in the sky. There are a strong permanent air streams at a high altitude at many regions of the USA.
8. The installation can be easy relocated in other place.

As with any new idea, the suggested concept is in need of research and development. The theoretical problems do not require fundamental breakthroughs. It is necessary to design small, free flying installations to study and get an experience in the design, launch, stability, and the cable energy transmission from a flying wind turbine to a ground electric generator.

This paper has suggested some design solutions from patent application [4]. The author has many detailed analysis in addition to these presented projects. Organizations interested in these projects can address the author (http://Bolonkin.narod.ru , aBolonkin@juno.com , abolonkin@gmail.com). The other ideas are in [11] - [50].

References

(Reader can find part of these articles in WEBs: http://Bolonkin.narod.ru/p65.htm, http://arxiv.org, search: Bolonkin, and in the book "*Non-Rocket Space Launch and Flight*", Elsevier, London, 2006, 488 pgs.)

[1] Bolonkin A.A., *Utilization of Wind Energy at High Altitude*, AIAA-2004-5756, AIAA-2004-5705. International Energy Conversion Engineering Conference at Providence, RI, USA, Aug.16-19, 2004.
[2] Gipe P., *Wind Power*, Chelsea Green Publishing Co., Vermont, 1998.
[3] Thresher R.W. and etc, *Wind Technology Development*: Large and Small Turbines, NRFL, 1999.
[4] Bolonkin, A.A., "*Method of Utilization a Flow Energy and Power Installation for It*", USA patent application 09/946,497 of 09/06/2001.

[5] Bolonkin, A.A., *Transmission Mechanical Energy to Long Distance*. AIAA-2004-5660.
[6] Galasso F.S., *Advanced Fibers and Composite,* Gordon and Branch Scientific Publisher, 1989.
[7] Carbon and High Performance Fibers Directory and Data Book, London-New. York: Chapmenand Hall, 1995, 6th ed., 385 p.
[8] Concise *Encyclopedia of Polymer Science and Engineering*, Ed. J.I.Kroschwitz, N. Y.,Wiley, 1990, 1341 p.
[9] Dresselhaus, M.S., *Carbon Nanotubes*, by, Springer, 2000.
[10] Bolonkin, A.A., "Inexpensive Cable Space Launcher of High Capability", IAC-02-V.P.07, 53rd International Astronautical Congress. The World Space Congress – 2002, 10-19 Oct. 2002/Houston, Texas, USA. *JBIS, Vol.56*, pp.394-404, 2003.
[11] Bolonkin, A.A, "Non-Rocket Missile Rope Launcher", IAC-02-IAA.S.P.14, 53rd International Astronautical Congress. The World Space Congress – 2002, 10-19 Oct 2002/Houston, Texas, USA. *JBIS, Vol.*56, pp.394-404, 2003.
[12] Bolonkin, A.A., "Hypersonic Launch System of Capability up 500 tons per day and Delivery Cost $1 per Lb". IAC-02-S.P.15, 53rd International Astronautical Congress. The World Space Congress – 2002, 10-19 Oct 2002/Houston, Texas, USA. *JBIS, Vol.57,* pp.162-172. 2004.
[13] Bolonkin, A.A., "Employment Asteroids for Movement of Space Ship and Probes". IAC-02-S.6.04, 53rd International Astronautical Congress. The World Space Congress – 2002, 10-19 Oct. 2002/Houston, USA. *JBIS, Vol.56*, pp.98-197, 2003.
[14] Bolonkin, A.A., "Non-Rocket Space Rope Launcher for People", IAC-02-V.P.06, 53rd International Astronautical Congress. The World Space Congress – 2002, 10-19 Oct 2002/Houston, Texas, USA. *JBIS, Vol.56*, pp.231-249, 2003.
[15] Bolonkin, A.A., "*Optimal Inflatable Space Towers of High Height*". COSPAR-02 C1.1-0035-02, 34th Scientific Assembly of the Committee on Space Research (COSPAR). The *World Space Congress – 2002, 10-19 Oct 2002/Houston, Texas, USA. JBIS*, Vol.56, pp.87-97, 2003.
[16] Bolonkin, A.A., "Non-Rocket Earth-Moon Transport System", COSPAR-02 B0.3-F3.3-0032-02, 34th Scientific Assembly of the Committee on Space Research (COSPAR). The World Space Congress – 2002, 10-19 Oct 2002, Houston, Texas, USA. "*Advanced Space Research"*, Vol.31, No. 11, pp. 2485-2490, 2003.
[17] Bolonkin, A.A., "Non-Rocket Earth-Mars Transport System", COSPAR-02B0.4-C3.4-0036-02, 34th Scientific Assembly of the Committee on Space Research (COSPAR). The World Space Congress – 2002, 10-19 Oct 2002/Houston, Texas, USA. *Actual problems of aviation and space system*. No.1(15), vol.8, pp.63-73, 2003.
[18] Bolonkin, A.A., "Transport System for delivery Tourists at Altitude 140 km". IAC-02-IAA.1.3.03, 53rd International Astronautical Congress. The World Space Congress – 2002, 10-19 Oct. 2002/Houston, Texas, USA. *JBIS, Vol.56*, pp.314-327, 2003.
[19] Bolonkin, A.A., "*Hypersonic Gas-Rocket Launch System.*", AIAA-2002-3927, 38th AIAA/ASME/SAE/ ASEE Joint Propulsion Conference and Exhibit, 7-10 July, 2002. Indianapolis, IN, USA.
[20] Bolonkin, A.A., Multi-Reflex Propulsion Systems for Space and Air Vehicles and Energy Transfer for Long Distance, *JBIS, Vol, 57*, pp.379-390, 2004.
[21] Bolonkin A.A., *Electrostatic Solar Wind Propulsion System*, AIAA-2005-3653. 41 Propulsion Conference, 10-12 July, 2005, Tucson, Arizona, USA.
[22] Bolonkin A.A., *Electrostatic Utilization of Asteroids for Space Flight*, AIAA-2005-4032. 41 Propulsion Conference, 10-12 July, 2005, Tucson, Arizona, USA.
[23] Bolonkin A.A., *Kinetic Anti-Gravitator*, AIAA-2005-4504. 41 Propulsion Conference, 10-12 July, 2005, Tucson, Arizona, USA.

[24] Bolonkin A.A., Sling Rotary Space Launcher, AIAA-2005-4035. 41 Propulsion Conference, 10-12 July, 2005, Tucson, Arizona, USA.
[25] Bolonkin A.A., *Radioisotope Space Sail and Electric Generator*, AIAA-2005-4225. 41 Propulsion Conference, 10-12 July, 2005, Tucson, Arizona, USA.
[26] Bolonkin A.A., *Guided Solar Sail and Electric Generator*, AIAA-2005-3857. 41 Propulsion Conference, 10-12 July, 2005, Tucson, Arizona, USA.
[27] Bolonkin A.A., *Problems of Electrostatic Levitation and Artificial Gravity*, AIAA-2005-4465. 41 Propulsion Conference, 10-12 July, 2005, Tucson, Arizona, USA.
[28] A.A. Bolonkin, *Space Propulsion using Solar Wing and Installation for It*. Russian patent application #3635955/23 126453, 19 August, 1983 (in Russian). Russian PTO.
[29] A, Bolonkin, Installation for Open Electrostatic Field. Russian patent application #3467270/21 116676, 9 July, 1982 (in Russian). Russian PTO.
[30] A.A.Bolonkin, *Getting of Electric Energy from Space and Installation for It*. Russian patent application #3638699/25 126303, 19 August, 1983 (in Russian). Russian PTO.
[31] A.A.Bolonkin, *Protection from Charged Particles in Space and Installation for It*. Russian patent application #3644168 136270 of 23 September 1983, (in Russian). Russian PTO.
[32] A.A.Bolonkin, *Method of Transformation of Plasma Energy in Electric Current and Installation for It*. Russian patent application #3647344 136681 of 27 July 1983 (in Russian), Russian PTO.
[33] A.A.Bolonkin, *Method of Propulsion using Radioisotope Energy and Installation for It. of Plasma Energy in Electric Current and Installation for it*. Russian patent application #3601164/25 086973 of 6 June, 1983 (in Russian), Russian PTO.
[34] A.A.Bolonkin, *Transformation of Energy of Rarefaction Plasma in Electric Current and Installation for it*. Russian patent application #3663911/25 159775 of 23 November 1983 (in Russian). Russian PTO.
[35] A.A.Bolonkin, *Method of a Keeping of a Neutral Plasma and Installation for it*. Russian patent application #3600272/25 086993 of 6 june 1983 (in Russian). Russian PTO.
[36] A.A.Bolonkin, *Radioisotope Propulsion. Russian patent application* #3467762/25 116952 of 9 July 1982 (in Russian). Russian PTO.
[37] A.A.Bolonkin, *Radioisotope Electric Generator*. Russian patent application #3469511/25 116927 of 9 July 1982 (in Russian). Russian PTO.
[38] A.A.Bolonkin, *Radioisotope Electric Generator*. Russian patent application #3620051/25 108943 of 13 July 1983 (in Russian). Russian PTO.
[39] A.A.Bolonkin, Method of Energy Transformation of Radioisotope Matter in Electricity and Installation for it. Russian patent application #3647343/25 136692 of 27 July 1983 (in Russian). Russian PTO.
[40] A.A.Bolonkin, *Method of stretching of thin film. Russian patent application* #3646689/10 138085 of 28 September 1983 (in Russian). Russian PTO.
[41] Bolonkin, A.A. and R.B. Cathcart, Inflatable 'Evergreen' Dome Settlements for Earth's Polar Regions. *Clean. Techn. Environ. Policy*. DOI 10.1007/s10098.006-0073.4 .
[42] Bolonkin, A.A. and R.B. Cathcart, "A Cable Space Transportation System at the Earth's Poles to Support Exploitation of the Moon", *Journal of the British Interplanetary Society* 59: 375-380, 2006.
[43] Bolonkin A.A., Cheap Textile Dam Protection of Seaport Cities against Hurricane Storm Surge Waves, Tsunamis, and Other Weather-Related Floods, 2006. http://arxiv.org.
[44] Bolonkin, A.A. and R.B. Cathcart, Antarctica: A Southern Hemisphere Windpower Station? Arxiv, 2007.
[45] Bolonkin A.A., Cathcart R.B., Inflatable 'Evergreen' Polar Zone Dome (EPZD) Settlements, 2006. http://arxiv.org

[46] Bolonkin, A.A. and R.B. Cathcart, The Java-Sumatra Aerial Mega-Tramway, 2006. http://arxiv.org.
[47] Bolonkin, A.A., "Optimal Inflatable Space Towers with 3-100 km Height", *Journal of the British Interplanetary Society* Vol. 56, pp. 87 - 97, 2003.
[48] Bolonkin A.A., *Non-Rocket Space Launch and Flight, Elsevier*, London, 2006, 488 pgs.
[49] Macro-Engineering: *A Challenge for the Future*. Springer, 2006. 318 ps. Collection of articles.
[50] Cathcart R.B. and Bolonkin, A.A. Ocean Terracing, 2006. http://arxiv.org.
2003

Article Airborne Wind Turbines for AEAT 4 24 13

Chapter 2.
Airborne Wind Turbines

Abstract (Review)

Propose – research and review of airborne wind turbines and transfer energy from turbine to dround surface.
Design/methodology/approach – consideration and comparison the different design of air borne turbines.
Findings – offered and comparison different turbines. R&D of their theory.
Research limitations/implications – transfer energy from airborn wind turbine to ground.

Practical implications – selection of the optimal air borne wind turbine and transwer energy. This article suggests a revolutionary innovation which produces a dramatic increase in power per unit at a lower cost per unit of energy extracted and is independent of prevailing weather. The main innovation consists of large free-flying air rotors positioned at high altitude for power and stable air stream, and two types (mechanical and electrical) of an energy cable transmission system between the air rotor and a ground system. The air rotor system flies at high altitude up to 10 km. Stability and control systems is provided which also enable changing altitude.

Originality/value – offered new turbines and transfer energy. Author also provides a brief review of main wind systems/turbines describing their advantages and disadvantages.

Keywords: wind energy, cable energy transmission, electric airborne transmission, utilization of wind energy at high altitude, air rotor, airborne wind turbines.

Introduction

High Altitude Winds.

Power generation from winds usually comes from winds very close to the surface of the earth. Winds at higher altitudes are stronger and more consistent, and may have a global capacity of 380 TW. Recent years have seen significant advances in technologies meant to generate electricity from high altitude winds. Worldwide there are now over two hundred thousand wind turbines operating, with a total nameplate capacity of 282,482 MW as of end 2012. The European Union alone passed some 100,000 MW nameplate capacities in September 2012, while the United States surpassed 50,000 MW in August 2012 and China passed 50,000 MW the same month.

Some Information about Wind Energy.

The power of wind engine strongly depends on wind speed (to the third power). Low altitude wind ($H = 10$ m) has the standard average speed of $V = 6$ m/s. High altitude wind is powerful and practically everywhere is stable and constant. Wind in the troposphere and stratosphere are powerful and permanent. For example, at an altitude of 5 km, the average wind speed is about 20 M/s, at an altitude 10 - 12 km the wind may reach 40 m/s (at latitude of about 20 - 35^0 N).

There are permanent jet streams at high altitude. For example, at $H = 12$-13 km and about 25^0 N latitude, the average wind speed at its core is about 148 km/h (41 m/s). The most intensive portion has a maximum speed of 185 km/h (51 m/s) latitude 22^0, and 151 km/h (42 m/s) at latitude 35^0 in North America. On a given winter day, speeds in the jet core may exceed 370 km/h (103 m/s) for a distance of several hundred miles along the direction of the wind. Lateral wind shears in the direction normal to the jet stream may be 185 km/h per 556 km to right and 185 km/h per 185 km to the left.

The wind speed of $V = 40$ m/s at an altitude $H = 13$ km provides 64 times more energy than surface wind speeds of 6 m/s at an altitude of 10 m.

This is an enormous renewable and free energy source. (See reference: *Science and Technology, v.2,* p.265).

High altitude jet stream.

Jet streams are fast flowing, narrow air currents found in the atmospheres of some planets, including Earth. The main jet streams are located near the tropopause, the transition between the troposphere (where temperature decreases with altitude) and the stratosphere (where temperature increases with altitude). The major jet streams on Earth are westerly winds (flowing west to east). Their paths typically have a meandering shape; jet streams may start, stop, split into two or more parts, combine into one stream, or flow in various directions including the opposite direction of most of the jet. The strongest jet streams are the **polar jets**, at around 7–12 km (23,000–39,000 ft.) above sea level, and the higher and somewhat weaker **subtropical jets** at around 10–16 km (33,000–52,000 ft.). The Northern Hemisphere and the Southern Hemisphere each have both a polar jet and a subtropical jet. The northern hemisphere polar jet flows over the middle to northern latitudes of North America, Europe, and Asia and their intervening oceans. The southern hemisphere polar jet mostly circles Antarctica all year round.

Jet streams are caused by a combination of a planet's rotation on its axis and atmospheric heating (by solar radiation and, on some planets other than Earth, internal heat). Jet streams form near boundaries of adjacent air masses with significant differences in temperature, such as the polar region and the warmer air towards the equator.

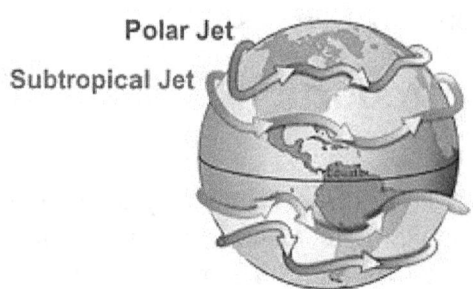

Figure 1. General configuration of the polar and subtropical jet streams.

Other jet streams also exist. During the northern hemisphere summer, easterly jets can form in tropical regions, typically in a region where dry air encounters more humid air at high altitudes. Low-level jets also are typical of various regions such as the central United States.

Meteorologists use the location of some of the jet streams as an aid in weather forecasting. The main commercial relevance of the jet streams is in air travel, as flight time can be dramatically affected by either flying with the flow or against the flow of a jet stream. Clear-air turbulence, a potential hazard to aircraft passenger safety, is often found in a jet stream's vicinity but does not create a substantial alteration on flight times.

Economy of conventional utilization wind energy.

Wind power plants have low ongoing costs, but moderate capital cost. The marginal cost of wind energy once a plant is constructed is usually less than 1-cent per kW·h. As wind turbine technology improved this cost has been reduced. There are now longer and lighter wind turbine blades (up 75 m), improvements in turbine performance and increased power generation efficiency. Also, wind project capital and maintenance costs have continued to decline.

The estimated average cost per unit incorporates the cost of construction of the turbine and transmission facilities, borrowed funds, return to investors (including cost of risk), estimated annual production, and other components, averaged over the projected useful life of the equipment, which may be in excess of twenty years. Energy cost estimates are highly dependent on these assumptions so published cost figures can differ substantially. In 2004, wind energy cost a fifth of what it did in the 1980s, and a continued downward trend is expected as larger multi-megawatt turbines were mass-produced. As of 2012 capital costs for wind turbines are substantially lower than 2008–2010 but are still above 2002 levels. A 2011 report from the American Wind Energy Association stated, "Wind's costs have dropped over the past two years, in the range of 5 to 6 cents per kilowatt-hour recently.... about 2 cents cheaper than coal-fired electricity, and more projects were financed through debt arrangements than tax equity structures last year.... winning more mainstream acceptance from Wall Street's banks.... Equipment makers can also deliver products in the same year that they are ordered instead of waiting up to three years as was the case in previous cycles.... 5,600 MW of new installed capacity is under construction in the United States, more than double the number at this point in 2010. Thirty-five percent of all new power generation built in the United States since 2005 has come from wind, more than new gas and coal plants combined, as power providers are increasingly enticed to wind energy as a convenient hedge against unpredictable commodity price moves."

A British Wind Energy Association report gives an average generation cost of onshore wind power of around 3.2 pence (between US 5 and 6 cents) per kW·h (2005). Cost per unit of energy produced was estimated in 2006 to be comparable to the cost of new generating capacity in the US for coal and natural gas: wind cost was estimated at $55.80 per MW·h, coal at $53.10/MW·h and natural gas at $52.50. Similar comparative results with natural gas were obtained in a governmental study in the UK in 2011. A 2009 study on wind power in Spain by Gabriel Calzada Alvarez of King Juan Carlos University concluded that each installed MW of wind power led to the loss of 4.27 jobs, by raising energy costs and driving away electricity-intensive businesses. The U.S. Department of Energy found the study to be seriously flawed, and the conclusion unsupported. The presence of wind energy, even when subsidized, can reduce costs for consumers (€5 billion/yr in Germany) by reducing the marginal price, by minimizing the use of expensive peaking power plants.

In February 2013 Bloomberg New Energy Finance reported that the cost of generating electricity from new wind farms is cheaper than new coal or new baseload gas plants. In Australia, when including the current Australian federal government carbon pricing scheme their modeling gives costs (in Australian dollars) of $80/MWh for new wind farms, $143/MWh for new coal plants and $116/MWh for new baseload gas plants. The modeling also shows that "even without a carbon price (the most efficient way to reduce economy-wide emissions) wind energy is 14% cheaper than new coal and 18% cheaper than new gas." Part of the higher costs for new coal plants is due to high financial lending costs because of "the reputational damage of emissions-intensive investments". The expense of gas fired plants is partly due to "export market" effects on local prices. Costs of production from coal fired plants built in "the 1970s and 1980s" are cheaper than renewable energy sources because of depreciation.

High-altitude wind power (HAWP)

HAWP has been imagined as a source of useful energy since 1833 with John Etzler's vision of capturing the power of winds high in the sky by use of tether and cable technology. An atlas of the high-altitude wind power resource has been prepared for all points on Earth. A similar atlas of global assessment was developed at Joby Energy. The results were presented at the first annual Airborne Wind Energy Conference held at Stanford University by Airborne Wind Energy Consortium.

Various mechanisms are proposed for capturing the kinetic energy of winds such as kites, kytoons, aerostats, gliders, gliders with turbines for regenerative soaring, sailplanes with turbines, or other airfoils, including multiple-point building- or terrain-enabled holdings. Once the mechanical energy is derived from the wind's kinetic energy, then many options are available for using that mechanical energy: direct traction, conversion to electricity aloft or at ground station, conversion to laser or microwave for power beaming to other aircraft or ground receivers. Energy generated by a high-altitude system may be used aloft or sent to the ground surface by conducting cables, mechanical force through a tether, rotation of endless line loop, movement of changed chemicals, flow of high-pressure gases, flow of low-pressure gases, or laser or microwave power beams. There are two major scientific articles about jet stream power.

Programs for Developing Wind Energy

Wind is a clean and inexhaustible source of energy that has been used for many centuries to grind grain, pump water, propel sailing ships, and perform other work. Wind farm is the term used for a large number of wind machines clustered at a site with persistent favorable winds, generally near mountain passes. Wind farms have been erected in New Hampshire, in the Tehachapi Mountains. at Altamont Pass in California, at various sites in Hawaii, and may other locations. Machine capacities range from 10 to 500 kilowatts. In 1984 the total energy output of all wind farms in the United States exceeded 150 million kilowatt-hours.

A program of the United States Department of Energy encouraged the development of new machines, the construction of wind farms, and an evaluation of the economic effect of large-scale use of wind power.

The utilization of renewable energy ('green' energy) is currently on the increase. For example, numerous wind turbines are being installed along the British coast. In addition, the British government has plans to develop off-shore wind farms along their coast in an attempt to increase the use of renewable energy sources. A total of $2.4 billion was injected into renewable energy projects over the last three years in an attempt to meet the government's target of using renewable energy to generate 10% of the country's energy needs by 2010.

This British program saves the emission of almost a million tons of carbon dioxide. Denmark plans to get about 30% of their energy from wind sources.

Unfortunately, current ground wind energy systems have deficiencies which limit their commercial applications:

1. Wind energy is unevenly distributed and has relatively low energy density. Huge turbines cannot be placed on the ground; many small turbines must be used instead. In California, there are thousands of small wind turbines. However, while small turbines are relatively inefficient, very huge turbines placed at ground are also inefficient due to the relatively low wind energy density and their high cost. The current cost of wind energy is higher than energy of thermal power stations.

2. Wind power is a function of the cube of wind velocity. At surface level, wind has low speed and it is non-steady. If wind velocity decreases in half, the wind power decreases by a factor of 8 times.

3. The productivity of a wind-power system depends heavily on the prevailing weather.

4. Wind turbines produce noise and visually detract from the landscape.

While there are many research programs and proposals for the wind driven power generation systems, all of them are ground or tower based. The system proposed in this article is located at high altitude (up to the stratosphere), where strong permanent and steady streams are located. This article also proposes a solution to the main technologist challenge of this system; the transfer of energy to the ground via a mechanical transmission made from closed loop, modern composite fiber cable.

The reader can find the information about this idea in [1]-[2], a detailed description of the innovation in [3]-[6], and the wind energy in references [7]-[8], new material used in the proposed innovation in [9]-[13]. The review of last airborne concepts in [14]-[17].

Innovation

Innovation wind turbines for high altitudes are presented in figures 2 – 5 (see figs. 1a – 5 in Chapter 1).

Figures 2-5 (see figs. 1a – 5 in Chapter 1).

Brief Theory of Estimation of Airborne Wind Installations

Wind (Speed, Duration, Altitude Distribution, Speed Distribution)

We can calculate the minimum and maximum acceptable wind necessary for operation of the air borne wind installation (ABWI). Our purpose is estimation of time (% or a number of days/hours in year) when the ABWI cannot operate.

Annual average wind speed. The United States Annual Average Wing Speed is taken from a map in **Wind Energy Resource Atlas of the United States**. The map was published in 1987 by Battelle's Pacific Northwest Laboratory for the U.S. Department of Energy. The complete atlas can be obtained by writing the American Wind Energy Association or the National Technical Information Service. The same maps are accessible around the world. They are presented in publication of the USA Department of Energy. The maps show the average wind speed at altitude 10 and 50 meters. This speed is 4 - 8 m/sec.

Wind speed and Height. Wind speed increases with height. The speed may be computed by equation

$$\frac{V}{V_0} = \left(\frac{H}{H_0}\right)^\alpha \qquad (1)$$

where V_0 is the wind speed at the original height, V the speed at the new height, H_0 the original height, H the new height, and α the surface roughness exponent (Table 2).

Table 2. Typical surface roughness exponents for power law method of estimating changes in wind speed with height

Terrain	Surface Roughness Exponent, α
Water or ice	0.10
Low grass or steppe	0.14
Rural with obstacles	0.20
Suburb and woodlands	0.25

Reference: P.Gipe, Wind Energy comes of Age, 1995, [7].

The result of computation of equation (1) for different α is presented at Figure 6. The wind speed increases on 20 - 50% with height 1000 m.

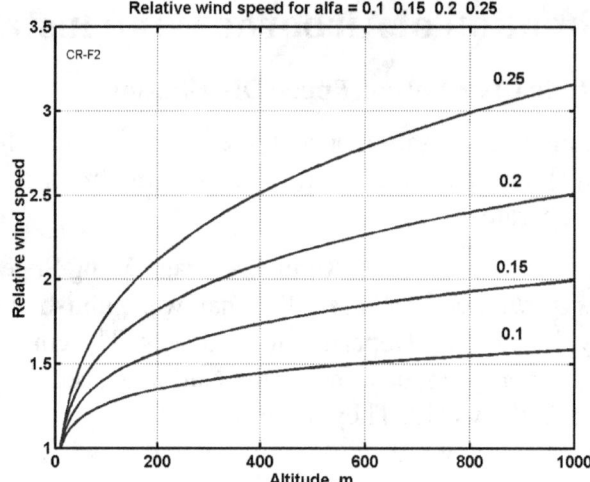

Figure 6. Relative wind speed via altitude and Earth surface. For sea and ice α = 0.1.

Annual Wind speed distribution. Annual speed distributions vary widely from one site to another, reflecting climatic and geographic conditions. Meteorologists have found that Weibull probability function best approximates the distribution of wind speeds over time at sites around the world where actual distributions of wind speeds are unavailable. The Rayleigh distribution is a special case of the Weibull function, requiring only the average speed to define the shape of the distribution.

Equation of Rayleigh distribution is

$$f_x(x) = \frac{x}{\alpha^2}\exp\left[-\frac{1}{2}\left(\frac{x}{\alpha}\right)^2\right], \quad x \geq 0, \quad E(X) = \sqrt{\frac{\pi}{2}}\alpha, \quad Var(X) = \left(2 - \frac{\pi}{2}\right)\alpha^2, \qquad (2)$$

where α is parameter.

Figure 7 presents the annual wind distribution of average speeds 4, 5, and 6 m/s. These data gives possibility to easy calculate the amount (percent) days (time) when ABWI can operate in year (Figure 8). It is very important value for the estimation efficiency of offered turbines.

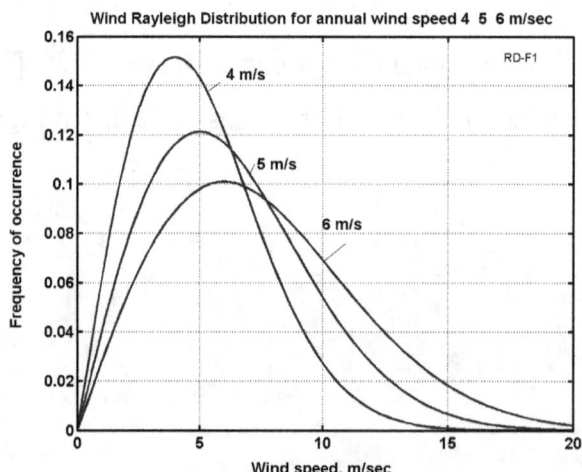

Figure 7. Wind speed distribution.

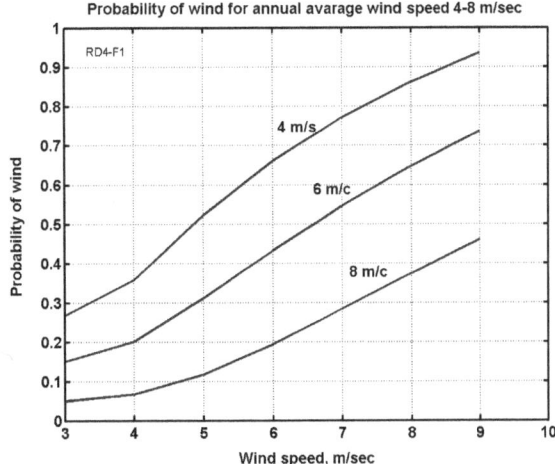

Figure 8. Probability of wind for annual average wind speed 4 - 8 m/s.

Let us compute two examples:

Assume, the observer has minimum wind speed 3 m/s, maximum safety speed 25 m/s, altitude 100 m, the average annual speed in given region is 6 m/s. From Figure 6, 7, 8, Eq. (1), we can get the wind speed is 8.4 at $H = 100$m, the probability that the wind speed will be less the 2 m/s is 8%, less 3 m/sec is 15%, the probability that the wind speed will be more 25 m/s is closed to 0.

Forces of the Airborne Wind Installation

The next forces are acting in airborne wind installation: lift forces of wing and dirigible (air balloon), weight of installation (turbine + electric generator and transformer), approximately half of main cable weight, approximately half of transmission weight, drag of turbine, drag of wing, drag of dirigible (if one is used), approximately half drag of main cable, approximately half drag of transmission cable.

These forces are presented in figure 9.

The balance equations in axis x (horizontal) and axis y (vertical) are:

$$\sum_x \quad F_c \cos \alpha = D_r + D_w + D_d + 0.5 D_c + 0.5 D_{tr}, \quad (3)$$
$$\sum_y \quad L_w + L_d = F_c \sin \alpha + Mg + 0.5 m_c g + 0.5 m_{tr} g. \quad (4)$$

Here F_c is force of main cable, N; D_r is air drag of wind rotor/turbine, N; D_w is air drag of wing, N; D_d is air drag of dirigible, N; D_c is air of main cable, N; D_{tr} is air drag of transmission, N; L_w is wing lift force, N; L_d is dirigible lift force, N; M is mass of installation (air turbine + electric generator and transformer), kg; $g = 9.81$ m/s is Earth gravity; m_c is mass of main cable, kg; m_{tr} is mass of transmission cable, kg; α is angle between line from initial point at Earth to air installation and Earth surface.

For given design parameters, given angle α ($\alpha \approx 25^\circ \div 35^\circ$) and the given row of the wind speed (from given V_{min} throw the safety V_{max}) we can find (after using the equation below) the cable force F_c from Eq. (3) and requested the wing force L_w from Eq. (4) and compare with initial data (cross section of main cable area). If they are significantly different – recalculate for new data.

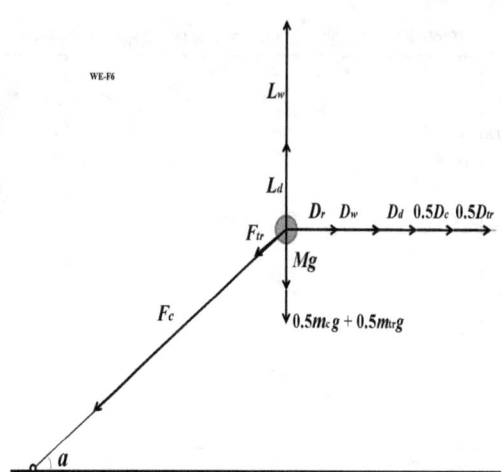

Figure 9. Forces active in air borne wind installation

Rotor Computation.
Power of a wind energy N [Watt, Joule/sec]
$$N = 0.5\,\eta\rho AV^3 \quad [W]\,. \tag{5}$$

The coefficient of efficiency, η, equals $0.15 \div 0.35$ for low speed propeller rotors (ratio of blade tip speed to wind speed equals $\lambda \approx 1$); $\eta = 0.45 \div 0.5$ for high speed propeller rotors ($\lambda = 5\text{-}7$). The Darrieus rotor has $\eta = 0.35 - 0.4$. The gyroplane rotor has $0.1 \div 0.15$. The air balloon and the drag (parachute) rotor has $\eta = 0.15 - 0.2$. The Makani rotor has $0.15 \div 0.25$. The theoretical maximum equals $\eta = 0.67$. A - front area of rotor, air balloon or parachute [m²]. ρ - density of air: $\rho_o = 1.225$ kg/m³ for air at sea level altitude $H = 0$; $\rho = 0.736$ at altitude $H = 5$ km; $\rho = 0.413$ at $H = 10$ km. V is average annually wind speed, m/s.

Table 3. Relative density ρ_r and temperature of the standard atmosphere via altitude

H, km	0	0.4	1	2	3	6	8	10	12
$\rho_r = \rho/\rho_o$	1	0.954	0.887	0.784	0.692	0.466	0.352	0.261	0.191
T, K	288	287	282	276	269	250	237	223	217

Issue [6].

The salient point here is that the wind power very strong depends from the wind speed (in third order!). If the wind speed increases by two times, the power increases by 8 times. If the wind speed increases 3 times, the wind power increases 27 times!

The wind speed increases in altitude and can reach in constant air stream at altitude $H = 5 - 7$ km up $V = 30 - 40$ m/s. At altitude the wind is more stable/constant which is one of the major advantages that the airborne wind rotor can has over ground wind rotor.

For comparison of different systems of wind rotors the engineers must make computations for average annual wind speed $V_0 = 6$ m/s and altitude $H_0 = 10$ m. For standard wind speed and altitude the wind power equals 66 W. The modern propeller wind turbines have diameter up $132 - 154$ m. For their comparison the engineers take the average standard the $H_0 = 50$ m and $V_0 = 10$ m/s. The power of the propeller turbine having rotor diameter 154 m reaches up 5.6 MW for standard conditions.

The energy, E, is produced in one year is (1 year $\approx 30.2 \times 10^6$ work sec) [J]
$$E = 3600 \times 24 \times 350 N \approx 30 \times 10^6 N,\ [J]. \tag{6}$$

The drag of the rotor equals
$$D_r = N/V, \quad [\text{N}]. \tag{7}$$

The drag of the dirigible is
$$D_d = 0.5 C_{D,d} \rho V^2 A_d, \quad [\text{N}], \tag{8}$$
$C_{D,d} \approx 0.01 \div 0.03$ is coefficient of air drag; A_d is cross section of dirigible $A_d = \pi d^2/4$, m^2.

The lift force of the wing,
$$L_w, \text{ is } L_w = 0.5 C_L \rho V^2 A_w, \quad [\text{N}], \tag{9}$$
where C_L is lift coefficient (maximum $C_L \approx 2 - 2.5$); A_w is area of the wing, m^2.

The drag of the wing is
$$D_w = 0.5 C_D \rho V^2 A_w, \quad [\text{N}], \tag{10}$$
where C_D is the drag coefficient ($C_D \approx 0.02 \div 0.2$).

The air drag, D_c, of main cable and air drag, D_{tr}, of the transmission cable is
$$D_c = 0.5 C_{d,c} \rho V^2 H d_c, \quad D_{tr} = 0.5 C_{d,r} V^2 H d_{tr}, \quad [\text{N}], \tag{11}$$
where $C_{d,c}$ - drag coefficient of main cable, $C_{d,c} \approx 0.05 - 0.15$; H is rotor altitude, m; d_c is diameter of the main cable, m. $C_{d,r}$ - drag coefficient of the transmission cable, $C_{d,r} \approx 0.05 - 0.15$; d_{tr} is diameter of the transmission cable, m. Only half of this drag must be added to the total drag of wind installation:
$$D \approx D_r + D_w + D_d + 0.5 D_c + 0.5 D_{tr}, \quad [\text{N}] \tag{12}$$

If the wind installation is supported by dirigible, the lift force and air drag of dirigible must be added to wing lift force (6) and total (9) of system. The useful lift force of dirigible is about 5 N/m^3 (0,5 kg/m^3) at $H = 0$ and zero at $H = 6$ km. Full lift force is:
$$L = L_w + L_d - Mg - 0.5g(m_c + m_{tr}), \quad [\text{N}]. \tag{13}$$

Here M is mass of installation (propeller + reducer + electro-generator + transformer), kg; $g = 9.81$ m/s^2 is Earth acceleration. Lift force of dirigible $L_d \approx 5 U_d$ [N], where U_d is dirigible volume, m^3.

The mass of main and transmission cable are:
$$m_c = \gamma_c S_c L, \quad m_{tr} = 2\gamma_{tr} S_{tr} L, \quad [\text{kg}], \tag{14}$$
where γ_c is specific weight/density of cables, kg/m^3, $\gamma_c \approx 1500 \div 1800$ kg/m^3; S_c is cross section area of cables, m^2; L is length of cable, m.

Required diameter of propeller for the power $P = 100$ kW and $V = 10$ m/s is 22.5 m; for speed $V = 15$ m/s diameter is 12.3 m.

The optimal speed of the parachute rotor equals $1/3 V$ and the theoretical maximum of efficiency coefficient is $\eta = 0.5$, real is 0.2.

The average angle α of connection line to horizon is
$$\sin \alpha \approx L/D, \tag{15}$$

The annual energy produced by the wind energy extraction installation equals
$$E = 8.33 N \quad [\text{kWh}]. \tag{16}$$

Cable Energy Transfer, Wing Area, and other Parameters

Cross-section area of the mechanical transmission cable,
$$S_t, \text{ is } S_t = N/v\sigma, \tag{17}$$

where N is transmission energy, W; v is speed of mechanical transmission, m/s; σ is safety stress of the mechanical transmission cable N/m², for good artificial fibers $\sigma \approx 50 \div 100$ N/mm² ($\sigma \approx (50 \div 100) \times 10^6$ N/m²). For long mechanical transmission $v \approx 50 \div 150$ m/s.

The cable force from wind turbine is
$$F_t = N/v, \qquad (18)$$
For example, if the transmission energy is $N = 100$ kW, speed of the mechanical transmission is $v = 50$ m/s, safety stress of artificial fiber is $\sigma = 100$ kg/mm² $= 10^9$ N/m², the cross-section area of the mechanical transmission cable is $s_m = 2$ mm² $= 2 \times 10^{-6}$ m². Diameter of the cable is $d = 1.6$ mm². $F_t = N/v = 10^5/50 = 2000$ N.

The air drag of transmission cable D_t, opposed the moving force is
$$D_t = 0.5 C_D \rho v^2 S_t, \qquad (19)$$
where $C_D \approx 0.008 \div 0.012$ is air drag coefficient; ρ is air density, kg/m³, S_t is surface area of cable, m². The surface area of double transmission cable is
$$S_t = \pi d^2 L_c / 2, \qquad (20)$$
where d is diameter of the cable, m; L_c is length of the cable, m.

The coefficient of transmission efficiency is
$$\eta = 1 - D_t/F_t, \qquad (21)$$
For our cable and $L_c = 1$ km $= 1000$ m, the $S_t = 10$ m². $N = 100$ kW, $F_t = 2000$ N and air drag $D_t = 150$ N (Newton)/km, coefficient efficiency is $\eta = 0.9625$ km⁻¹.

Cross-section area of main cable, S_m, is
$$S_m = \frac{\sqrt{D^2 + L^2}}{\sigma} \;[m^2], \qquad (22)$$
where σ is the safety stress of the main cable N/m².

The production cost, c, in kWh is
$$c = \frac{M_0 + I_0/K_1}{E}, \qquad (23)$$
where M_0 – annual maintenance [\$]; I_0 - cost of Installation [\$]; K_1 - life time (years); E - annual energy produced by flow installation [J];

The annual profit
$$F_0 = (C-c)E. \qquad (24)$$
where F_0 – annual profit [\$]; C - retail price of 1 kWh [\$].

In first estimation of the required area of the support wing is about
$$A_w \approx \eta A \sin\alpha / C_L, \qquad (25)$$
where α is the angle between the support cable and horizontal surface.

The wing area is served by ailerons for balancing of the rotor (propeller) torque moment
$$A_a = \frac{\eta A R}{\lambda_i \Delta C_{L,a} r}, \qquad (26)$$
r - distance from center of wing to center of aileron [m]; R - radius of rotor (turbine)[m]; $\Delta C_{L,a}$ - difference of lift coefficient between left and right ailerons;

The minimum wind speed for installation support by the wing alone

$$V_{min} = \sqrt{\frac{2W}{C_{L,max}\rho A_w}}, \qquad (27)$$

where $W = L$ is force of the total weight of the airborne system including transmission, [N]. If a propeller rotor is used in a gyroplane mode, minimal speed will decrease by 2 – 2.5 times. If wind speed equals zero, the required power for driving the propeller in a propulsion (helicopter) mode is

$$N_s = W/K_2 \qquad [kW], \qquad (28)$$

where W - weight of installation (rotor + generator + transformer + cables)[kg]; K_2 – rotor lift coefficient (5 - 12 [kg/kW]).

The specific weight of energy storage (flywheel) can be estimated by

$$E_s = \sigma/2\gamma \qquad [J/kg]. \qquad (29)$$

For example, if $\sigma = 200$ kg/mm², $\gamma = 1800$ kg/m³, then $E_s = 0.56$ MJ/*kg* or $E_s = 0.15$ kWh/kg.

Electric Transfer of Energy

Properties of the matter needed for computation of characteristics of the electric line from airborne rotor to ground installation is below.

1. Electric current safety for different wires.

Table 4. Safe electric currents via different materials and cross-section of wires [16] p.115.

Cross-section wire mm²/matter	1	1.5	2.5	4	10	25	Resistance, Ohm·m ρ, 10^{-8}	Specific weight, γ, kg/m³
Aluminum	8	11	16	20	34	80	2.8	2700
Copper	11	14	20	25	43	100	1.75	8930
Iron	-	-	6	10	17	-	9.8	7900

Author employs electric wire design which allows permanently maintaining the electric current safely at about 10 A/mm². It is that value which we use in our calculation.

Table 5. Spark gap between bare wires in atmosphere. [16] p.126.

El.Voltage, kV	Distance, mm	El.Voltage, kV	Distance, mm	El.Voltage, kV	Distance, mm
20	16	100	200	300	600
40	46	200	410		

Table 6. Dielectric strength of insulators [5]-[6].

Matter	MV/m
Lexan	320 - 640

Kapton H	120 – 320
Mylar	160 - 640
Parylene	240 - 400
Polyethylene	500–700*
Vacuum	100
Air	1 - 3

*For room temperature

2. Mass m_e [kg/kW.km] of the 1 km electric wires is

$$s = P/(pU), \quad m_e = 2k_2\gamma sL, \qquad (30)$$

were s is cross section of electric wire, m²; $p \approx 5 \div 10$ A/mm² is safety electric current A/m²; U is voltage, V; $k_2 \approx 2 \div 3$ is insulator coefficient, γ is the specific weight of wire, kg/m³; L is length of wire, m; P is electric power, W. For example, if $P = 10^5$ W = 100 kW, $U = 10^4$ V, $p = 10$ A/mm² = 10^7 A/m², $\gamma = 2800$ kg/m³ (aluminum wire), $L = 1000$ m, the $s = 1$ mm², than $m_c \approx 11$ kg/km, or $m_c \approx 0.11$ kg/(kW.km).

3. Electric resistance and coefficient of electric efficiency are:

$$R = \rho_e L/s, \quad \eta = 1 - \Delta U/U = 1 - 2I\rho_e L/sU, \qquad (31)$$

where R is electric resistance, Ω; ρ_e is coefficient of electric resistance (Table # 1), ohm.m; η is coefficient electric efficiency; I is electric currency, A; ΔU is the loss of voltage in transmission wire, V; s is cross-section of wire, m². Example, if $P = 10^5$ W = 100 kW, $U = 10^4$ V, $p = 10$ A/mm² = 10^7 A/m², $\rho_e = 2.8 \times 10^{-8}$ Ω.m (aluminum wire), $L = 1000$ m, the $s = 1$ mm², then $\eta = 0.944$ km^{-1}.

4. Air drag of main cable and electric wires, connected in one cable is

$$D_{c+w} = 0.5 C_D \rho_a V^2 A_{c+w}, \quad [N], \quad A_{c+w} = s_{c+w} H, \qquad (32)$$

where C_D is the drag coefficient $C_D = 0.015 \div 0.15$; ρ_a – air density, $\rho_a \approx 1$ kg/m³; s_{c+w} is cross-section area of common cable, H is altitude, m. Example, if $s_{c+w} = 3 \times 10^{-6}$ m², $H = 1000$ m, $V = 15$ m/s, $C_D = 0.02$, then $D_{c+w} = 500$ N/km.

5. Electric generator.

Specific mass of the conventional (car) electric generator is about 4 – 5 kg/kW. This mass is inversely related to electric frequency. Standard electric frequency is 50 Hertz. Aviation generator which has frequency 400 Hertz has specific mass of about 0.5 kg/kW. Example, the aviation electric generator ГТ120 П46А (Russia) has power $N = 120$ kW, $U = 120/208$ V, frequency is $v = 400$ Hertz, $n = 100 \div 6000$ revolution/min, mass 67 kg, cooling by air. That means we can take for our estimation the specific weight about 0.5 kg/kW.

6. Transformer.

For passing the electric energy from airborne turbine to the Earth we need the electric transformer which converts the electric energy to high voltage. That allows decreasing the weight the electric wire.

The typical data of the conventional 3-fases transformer is following: the transformer having power 100 kW, frequency 50 Hertz has weight 505 kg, site 890×1105×600 mm, enter 400 V, exit 6/10 kV. The Transformer ТМГ-1000/6-10 has power 1000 kW, weight 2900 kg, frequency 50 Hertz, enter 400 V, exit 6/10 kV, cooling – oil. That is not suitable for us because the weight and size is big. If we will use the frequency 400 Hertz the transformer weight decreases in 400/50 = 8 times and equals about 0.5 kg/kW. That is acceptable. But it is possible that there will be cooling problem of generator and transformer.

The offered electric system needs in the frequency convector 400 Hertz to 50 Hertz or rectifier. But one is located on Earth surface and is needed for all airborne turbines having the electric transmission.

The total mass of electric transmission system (electric generator + transformer + wires) is about additional 1.2 ÷ 1.5 kg/kW in comparison with mechanical system having 0.3 ÷ 0.5 kg/kW. That also increases also the requested the wing area and weight, because the wing must support the full installation in minimal wind speed. But the electric translation system is better equipped for changing the altitude which allows selection of the altitude where is the wind speed is optimal. If we want an airborne wind system without transformer, we must design special high voltage generator.

The ABWI having an electric transmission is a high altitude lighting conductor in storm and, as such, is in need of special equipment for this case as protection or landing system.

7. Result of estimation the electric transfer/system.

The total mass of the airborne wind installation ($P = 100$ KW, $L = 1$ km) with electric transfer is:
Rotor (propeller): 1 kg/kW,
Wing: 1 ÷ 2 kg/m², or 1.5 ÷ 3 kg/kW,
Electric generator + transformer: 1 ÷ 1.2 kg/kW,
Main cable: 0.4 ÷ 0.6 kg/kW.km (turbine gets ≈ 50% of this weight),
Electric wires: 0.1 ÷ 0.15 kg/kW.km (turbine gets ≈ 50% of this weight); or
Mechanical transmission 0.1 ÷ 0.15 kg/kW.km (turbine gets ≈ 50% of this weight).
Total mass is about 4 ÷ 5 kg/kW, or 400 ÷ 500 kg (for average $P = 100$ kW). Mass of wing is 200 ÷ 250 kg (wing have the area 150 ÷ 200 m² and support the installation for a minimal wind speed 3 ÷ 5 m/s).

If airborne wind installation has the mechanical transmission then the total mass of installation will be about two times less, but airborne wind installation will require developing a special system for change the altitude.

The dirigible (special air balloon) can support the airborne in windless conditions. The needed volume is about 900 m³ for the electric transmission and 500 m³ for the mechanical transmission. Size of dirigible is 14×60 m and 10×45 m respectively. Support by dirigible is very useful because for exploitation of the airborne wind installation because we not expend energy for supporting the turbine at altitude in weak winds (speed less 3 m/s) or in windless conditions. This situation may be in 5 ÷ 10% of total time in low (< 1 ÷ 2 km) altitudes.

8. Electrostatic generator.

Electrostatic generator produces electricity of a very high voltage and is not encumbered by have heavy iron and wire, nor does it have a cooling problem. The relative mass may be less than mass of the magnetic generator and transformer. The estimation of mass can be made by equations: $m_g = M_g/P$, $P = IV$, $I = qv$, $q = cU$, $c = \varepsilon_o S/a$, (33)
where m_g is relative mass of electrostatic generator, kg/kW; M_g is mass of generator, kg; P is power, kW; I is electric currency, A; V is voltage, produced by generator, V; q is electric charge, C; v is

relative speed of generator plates, m/s; c is electric capacity of plates, F; U is voltage between plates, V; $\varepsilon_o = 8.85 \times 10^{-12}$ is electric constant, F/m; S is area of plates, m^2; a is distance between plates, m.

Let us, for example, take 250 plates of area 1 m^2 each with distance 2 mm and voltage between plates $U = 10^5$ V and thickness of isolator 1 mm, the plate speed $v = 700$ m/s. We take the exit voltage of generator $V = 2 \times 10^5$ V. Produced voltage V may be any (up 1 MV), but transfer more high voltage to Earth surface is difficult. Estimation show: the electric current may be $I = 350$ A and mass of generator $M_g = 1000$ kg, size 1.2×1.2 m (diameter × length). The produced energy will be $P = 70$ MW. The relative mass is $m_g \approx 0.015$ kg/kW which is a very small value which shows the electrostatic generator/engine is very perspective for R&D. But design power electrostatic generator is not an easy problem to solve.

Total Estimation and Optimization Airborne Wind System

Below are summary equations which help estimate and select the suitable parameters of installation. The first equation is preliminary; the second/last equation is final.

1. Relative mass m_e [kg/W] of the electric cable $m_e = M_e/N$, $m_e = 2k_1\gamma_e L/pU$, (34)
   where M_e is wire mass, kg; N is transfer power, W; $k_1 \approx 2$ is relative mass of insulator; γ_e is specific mass of wire, kg/m^3; L is length of wire, m; p is safety density of electric current, A/m^2; U is electric voltage of system, V.

2. Coefficient of electric efficiency of electric wire transmission

$$\eta = 1 - \Delta U/U, \quad \eta = 1 - 2p\rho L/U, \qquad (35)$$

   where ΔU is loss of voltage in transmission wire, V; U is voltage of full system V; ρ is specific electric resistance of wire, Ω.m. Increasing of voltage reduces the electric loss and mass of electric wire.

3. Relative mass m_g [kg/W] of the electric generator and electric transformer

$$m_g = 2k_2\mu_0\gamma/B^2v, \qquad (36)$$

   where $k_2 \approx 2$ is relative mass of generator/transformer wire; $\mu_o = 4\pi \times 10^{-7}$ is magnetic constant; $\gamma = 7900$ kg/m^3 is specific mass of the generator/transformer iron, $B \approx 1$ is maximal magnetic inductivity; v is electric frequency, Hertz. Increasing of the electric frequency reduces the generator and transformer mass, but complicates their cooling.

4. Relative mass m_c [kg/W] of main cable $m_c = M_c/N$, $m_c = 2\gamma_c L\cos\alpha/\sigma V$, (37)
   where M_c is mass of main cable, kg; σ is safety stress of main cable, N/m^2; V is wind speed, m/s; γ_c is the specific mass of the main cable;

5. Relative mass m_c [kg/W] of mechanical transmission cable
   $m_t = M_t/N$, $m_t = \gamma_t L/\sigma V$, (38)

   where M_t is mass of transmission cable, kg; σ is safety stress of transmission cable, N/m^2; V is wind speed, m/s; γ_t is the specific mass of the transmission cable.

6. Coefficient of efficiency the mechanical transmission

$$\eta = 1 - Dv/N, \quad \eta = 1 - C_f\rho v^3 Ld/N, \quad \eta = 1 - 2\pi^{0.5}C_f\rho v^3 L/(\sigma V N)^{0.5}, \qquad (39)$$

where D_f is friction drag of transmission, N; v is transmission speed, m/s; C_f is coefficient of friction drag; d is diameter of transmission cable, m. As you see the degreasing of the transmission speed v can significantly reduce the transmission loss. ρ is air density, kg/m^3.

Cost of construction and economy of wind turbines.
Cost of renewable energy

Average cost of the ground wind installation in 2012 were: 1 kW - $2K, 2 kW - $3.5K, 5 kW - $14K, 10kW – 35 ÷ 50K. Wind turbine $1,3 ÷ 2,2M per MW. Ground transmission $1500/km. The average allocation of cost: tower 27%, rotor blades 21%, generator 4%, transformer 4%, power convertor 6%, gearbox 11%, others 27%.

Table 7: Comparison of capital cost breakdown for typical onshore and offshore wind power systems in developed countries, 2011

Source: Blanco, 2009; EWEA, 2009; Douglas-Westwood, 2010; and Make Consulting, 2011c.

	Onshore	Offshore
Capital investment costs (USD/kW)	1 700-2 450	3 300-5 000
Wind turbine cost share[1] (%)	65-84	30-50
Grid connection cost share[2] (%)	9-14	15-30
Construction cost share[3] (%)	4-16	15-25
Other capital cost share[4] (%)	4-10	8-30

[1] *Wind turbine costs includes the turbine production, transportation and installation of the turbine.*
[2] *Grid connection costs include cabling, substations and buildings.*
[3] *The construction costs include transportation and installation of wind turbine and tower, construction wind turbine foundation (tower), and building roads and other related infrastructure required for installation of wind turbines.*
[4] *Other capital cost here include development and engineering costs, licensing procedures, consultancy and permits, SCADA (Supervisory, Control and Data Acquisition) and monitoring systems.*

Comparison of different airborne designs

There are a number of alternative designs of airborne wind turbines. Unfortunately in many cases the inventors are people who do not have the needed technical education, cannot develop the corresponded theory, and make the correct estimations and computations. Unfortunately, the entire wind energy industry is plagued by the paucity of contiguity of scientific knowhow and business acumen. Governmental agency and business leaders most often do not select the projects that are scientifically feasible. Conversely, some inventors are well connected with funding sources; be they governmental authorities or heads of large companies. They may receive large grants for perspective projects with little scientific merit. Before funding a high altitude wind energy device, mathematical modeling is necessary to detail the physics in order to persuade the experts that it is not only physically feasible but economically feasible and largely profitable.

Wind at high altitudes is faster and more consistent than winds near the Earth's surface and contains more than three times the power providing a phenomenal untapped resource. A comprehensive understanding of winds ranging from the upper boundary layer through the upper troposphere and its availability is critical to the development of our technology. Let us estimate the parameters of some airborne wind systems same power (100 kW). The first systems will have this power.

1. **Mogenn and system is lighter than air (MARS).**

Some of these systems shown in Fig. 10 are air balloon having shoulder blades which rotate the balloon under wind.

If the strong wind is $V = 15$ m/s and coefficient of efficiency $\eta = 0.15$ the requested the frond area of balloon is

$$A = P/(0.5\eta\rho V^3) \approx 400 \text{ m}^2 , \qquad (38)$$

If length of balloon is 3 times of diameter, the diameter of balloon will be about 12 m, length 36 m and volume 4500 m³. The helium cost was $\approx$ \$16/m³ at 2012. Total cost only helium is \$72K. Useful (without weight of balloon) lift force is 23000 N = 2300 kg. The mass of good generator + transmission $\approx$ 300 kg.

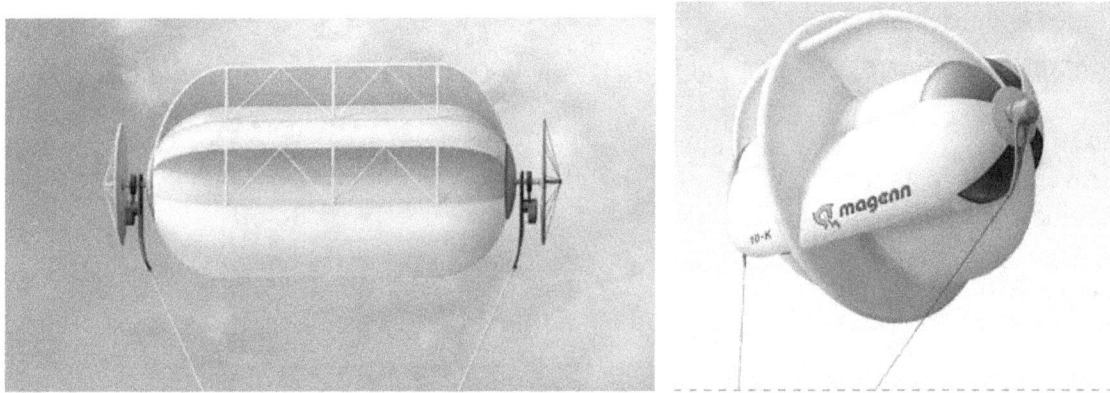

Fig.10. The airborne wind systems which are lighter than air.

Air drag of balloon is
$$D = 0.5C_D\rho V^2 A = 0.5 \times 0.3 \times 1 \times 15^2 \times 400 = 13500 \; N . \qquad (39)$$

Angle of the main cable to horizon in wind 15 m/s is about $35 \div 40°$. It is acceptable. But in storm the wind can reach the speed up 35 m/s and angle will be about $10 \div 12°$. That is not good especially at city having high buildings.

Magenn Power is developing a 10 kW airborne wind turbine system that floats 1,000 feet in the air, tethered to the ground. The inflatable Helium balloon portion of the device has vanes on it that capture the wind energy, similar to a paddle wheel, turning it on a horizontal axis that is fastened on two ends. A generator is affixed to both ends, and the electricity is transmitted down the tether to the ground.

The set-up costs for MARS are projected to run around \$4 to \$5 per Watt. In comparison, the set-up costs for a traditional utility-scale wind farm run around \$2.5 to \$3 per Watt. But those are huge installations, and require a good ground-level wind profile. The Magenn system can go where the wind farms are not feasible. The installation costs for a comparable Diesel generator system are about \$1.00 per Watt, but then there is the continual cost of the fuel to run the generators. Magenn has secured around a \$1 million (Canadian) grant from the Canadian government to further their refinement of the design. The grant is a matching-funds grant, contingent on Magenn being able to raise \$2 million from private sources. Magenn landed a separate \$300,000 grant to build a 1 kW sized unit. It is unknown what was actually built.

2. **The airborne wind propeller supported by dirigible.**

This design is presented in fig. 11. It is acceptable for altitude up 3 – 5 km. One may be also used for lifting and delivering of loads. Disadvantage is high cost of installation.

Fig. 11. Air borne wind propeller supported by dirigible.
The properties and data of this ABWI can be easily estimated by our theory.

3. The autogyro (gyroplane) rotor

Fig. 12 illustrates one of the autogyro designs by inventor Roberts.

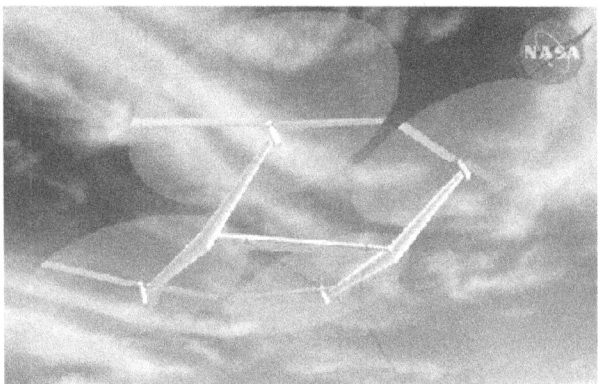

Fig. 12. Autogyro wind rotor

In Roberts design, if the wind is $V = 15$ m/s and coefficient of efficiency $\eta = 0.15$, the requested the area of propellers is $A = 400$ m^2. Or diameter of the 4 propellers is $D = 11.3$ m each. Gyroplanes rotor is easy for design. The flying windmills would initially get in position under their own power, using their motors to drive the propeller blades and helicopter upwards until they reached altitude. Then the motors would turn off and become generators as wind pushes the propeller blades, and the whirligig would float instead of fall because when tethered, the lift generated by the wind would overcome the craft's weight as it also generates power.

His claims are unrealistic because the power 240 kW for a diameter less than 10.7 m because the autogyro rotor axis has small angle to vertical line (10 ÷ 20° not 90° as conventional wind propeller). It is necessary because the autogyro rotor must also produce the vertical force for supporting the weight of installation. The problem of transporting of wind energy to the Earth surface is the same problem for all airborne wind rotors. The other problem is saving the installation in stormy weather because the propellers may be damaged by very strong wind. In contrast, the designs detailed in this paper include proposals which avoid these disadvantages.

4. Tube Airborne Wind Energy.

An air balloon tube and propeller installed inside tube is a wind installation is shown in fig. 13.

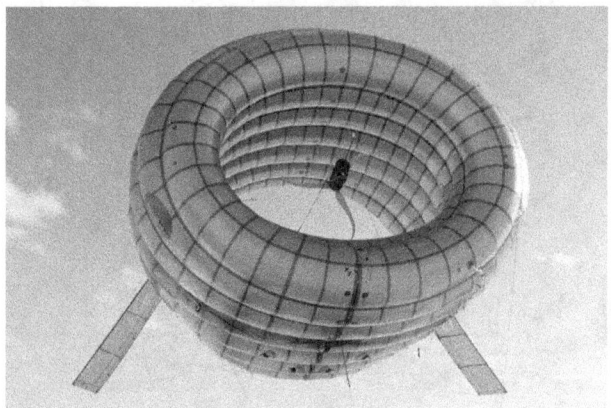

Fig.13. Altaeros Wind turbines

The company describes the installation as the Altaerod Airborne Wind Turbine, which makes use of an inflatable shell filled with helium, allowing it to gain high altitude. This gives it better access to more consistent and stronger winds, much higher than those turbines mounted on towers. The power uses tethers to reach the ground. Harnessing winds at higher altitudes will allow the turbine to reduce the costs of energy by almost 65%. Since it has a unique design that is easily installed, the start-up time amounts to only days, which means that each shell can be prepared and assembled more readily, for increased energy production."

The corrected design of tube can increase the speed inside maximum in $2^{0.5}$ times or the power in 2.8 times. But from figure 13 it is obvious that the inventor does not know aerodynamics and the presented installation is not efficient.

5. Makani Airborne Wind Turbine.

The original airborne turbine was offered by Makani figs.14 – 15. That is a single blade which flies in a circle. Blade has the propellers which produce the electric energy. If no wind the propellers may consume the energy from ground installation. They also lift the installation to altitude.

Fig. 14. Makani airborne wind turbine at Earth surface.

Fig.15. Makani Airborne Wind Turbine in air.

Joby Energy Co. is developing airborne wind turbines which will operate in the upper boundary layer and the upper troposphere. Their description from their company advertising: "Joby Energy's multi-wing structure supports an array of turbines. The turbines connect to motor-generators which produce thrust during takeoff and generate power during crosswind flight. Orientation in flight is maintained by an advanced computer system that drives aerodynamic surfaces on the wings and differentially controls rotor speeds. A reinforced composite tether transmits electricity and moors the system to the ground. The high redundancy of the array configuration can handle multiple points of failure and remain airborne. For launch, the turbines are supplied with power to enable vertical take-off. Upon reaching operating altitude, the system uses the power of the wind to fly cross-wind in a circular path. The high cross-wind speeds result in the turbines spinning the generators at high speeds, eliminating the need for gearboxes and increasing efficiency. The energy is transferred to the ground through the electrical tether. During occasional periods of low wind the turbines are powered to land the system safely."

Capacity Comparison. A comparison between the energy output potential of a 2 MW conventional turbine operating at 400 feet and a 2 MW Joby Energy airborne wind turbine operating at 2,000 feet shows a significant improvement in capacity factor. Our airborne wind turbine yields a capacity factor of nearly double the conventional turbine.

An airborne wind turbine must utilize less material than those found in ground based wind turbines. It is estimated that the Makani turbine will be 1/10 the weight of a standard wind turbine and cost half the price to install. It will be rated at the same amount of power. The price per kilowatt-hour would be even lower than coal-fired power at the present time, or about three cents per kilowatt hour.

The rotors on the flying wing of the Makani turbine function as generators and propellers. They use stored or backup power to reach their cruising altitude. When they reach 1,000 feet ($\approx$ 300 m) in altitude, they begin creating resistance to the higher winds and then generate electricity just like electric cars do with their brakes.

Is this turbine affected when there is no wind? The wing structures can use steady breezes to remain aloft, but if the wind goes below nine miles per hour, they would actually use electricity instead of generating it. Plans are to land the wing if there are long periods of forecasted low winds. But it will still be able to generate electricity with double the consistency of wind farms that are in operation today. This is due to the winds at the increased altitude, which may be twice as strong as those on the ground.

The future of the Makani airborne wind turbine looks quite bright. It won Popular Mechanics' Energy Breakthrough Award and got three million dollars in grant money from the Department of Energy. It also received 20 million dollars from Google, for venture capital funding.

In order to be fully successful, the airborne wind turbine must be able to generate a consistent and high rate of power. They are developing a larger turbine system that will float at about 1600 feet (≈ 500 m) in altitude, and this can potentially produce enough power for 600 houses. The prototype of this design should be launched in 2013 and in operation commercially in 2015. The Makani turbine may also be used above deeper offshore water, where even more energy can be produced. Fig.14 shows the company does not have good specialists. The offered installation is unstable and very complex in operation. Company received large sums of money but did not create any successful design.

Projects with mechanical transmission

Project 1. High-speed air propeller rotor (fig.2)

For example, let us consider a rotor diameter of 100 m (A = 7850 m^2), at an altitude H = 10 km (ρ = 0.4135 kg/m^3), wind speed of V = 30 m/s, an efficiency coefficient of η = 0.5, and a cable tensile stress of σ = 200 kg/mm^2. Then the power produced is N = 22 MW [Eq. (5)], which is sufficient for city with a population of 250,000. The rotor drag is D_r = 73 tons [Eq.(7)], the cross-section of the main cable area is S = $1.4D_r/\sigma$ =1.35×73/0.2 ≈ 500 mm^2, the cable diameter equals d = 25 mm; and the cable weight is W = 22.5 tons (for L = 25 km). The cross-section of the transmission cable is 36.5 mm^2, d = 6.8 mm, weight of two transmission cables is 3.33 tons for cable speed v =300 m/s [Eq.(14)]. The required wing size is 20×100 m (C_L = 0.8), wing area served by ailerons is 820 sq.m. If C_L=2, the minimum speed is 3 m/s. The installation will produce an annual energy E =190 GWh [Eq.(16)]. If the installation cost is $200K, has a useful life of 10 years, and requires maintenance of $50K per year, the production cost is c = 0.37 cent per kWh [Eq(23)]. If retail price is $0.15 per kWh, profit $0.1 per kWh, the total annual profit is $19 million per year [Eq.(24)].

Project 2. Air low speed wind engine with free flying cable flexible rotor (fig.3)

Let us consider the size of cable rotor of width 50 m, a rotor diameter of 1000 m, then the rotor area is A = 50×1000 = 50,000 sq.m. The angle rope to a horizon is 70°. The angle of ratio lift/drag is about 2.5°. The average conventional wind speed at an altitude H = 10 m is V = 6 m/s. It means that the speed at the altitude 1000 m is 11.4 - 15 m/s. Let us take average wind speed V=13 m/s at an altitude H = 1 km. The power of flow is N=$0.5 \cdot \rho V^3 A \cos 20^0$ =0.5×1.225×13^3×1000×50×0.94=63 MW.

If the coefficient efficiency is η = 0.2 the power of installation is η = 0.2×63 = 12.5 MW. The energy 12.5 MW is enough for a city with a population at 150,000. If we decrease our Installation to a 100x2000 m the power decreases approximately by 6 times (because the area decreases by 4 times, wind speed reaches more 15 m/s at this altitude. Power will be 75 MW. This is enough for a city with a population about 1 million of people.

If the average wind speed is different for given location the power for the basis installation will be: V =5 m/s, N=7.25 MW; V =6 m/s, N=12.5 MW; V=7 m/s, N= 19.9 MW; V = 8 m/s, N = 29,6 MW; V = 9 m/s, N = 42.2 MW; V = 10 m/s, N = 57.9 MW.

Economic efficiency

Let us assume that the cost of our installation is $1 million. According to the book "Wind Power" by P. Gipe [7], the conventional wind installation with the rotor diameter 7 m costs $20,000 and for average wind speeds of 6 m/s has power 2.28 kW, producing 20,000 kWh per year. To produce the same amount of power as our installation using by conventional methods, we would need 5482 (12500/2.28) conventional rotors, costing $110 million or 28M for costing 5K each installation. Let us assume that our installation has a useful life of 10 years and a maintenance cost is $50,000/year. Our installation produces 109,500,000 kWh energy per year. Production costs of energy will be approximately 150,000/109,500,000 = 0.14 cent/kWh. The retail price of 1 kWh of energy in New

York City is $0.15 now (2000). The revenue is 16 million dollars. If profit from 1 kWh is $0.1, the total profit is more 10 million dollars per year.

Estimation of some technical parameters.

The cross-section of main cable for an admissible fiber tensile strange σ= 200kg/sq.mm is S =2000/0.2 = 10,000 mm². That is two cables of diameter d = 80 mm. The weight of the cable for density 1800 kg/m³ is $W = SL\gamma = 0.01 \times 2000 \times 1800 = 36$ tons.

Let us assume that the weight of 1 sq.m of blade is 0.2 kg/m² and the weight of 1 m of bulk is 2 kg. The weight of the 1 blade will be 0.2 x 500 = 100 kg, and 200 blades are 20 tons. If the weight of one bulk is 0.1 ton, the weight of 200 bulks is 20 tons.

The total weight of main parts of the installation will be 94 tons. We assume 100 tons for purposes of our calculations.

The minimum wind speed when the flying rotor can supported in the air is (for C_y = 2)

$$V=(2Wg/C_L \rho S)^{0.5}=(2\times 100\times 10^4/2\times 1.225\times 200\times 500)^{0.5} = 2.86 \text{ m/s}$$

The probability of the wind speed falling below 3 m/s when the average speed is 12 m/s, is zero, and for 10 m/s is 0.0003. This equals 2.5 hours in one year, or less than one time per year. The wind at high altitude has greater speed and stability than near ground surface. There is a strong wind at high altitude even when wind near the ground is absent. This can be seen when the clouds move in a sky on a calm day.

Project 3. Low speed air drag rotor (fig.4)

Let us consider a parachute with a diameter of 100 m, length of rope 1500 m, distance between the parachutes 300 m, number of parachute 3000/300 = 10, number of worked parachute 5, the area of one parachute is 7850 sq.m, the total work area is A = 5 x 7850 = 3925 sq.m. The full power of the flow is 5.3 MW for V=6 m/s. If coefficient of efficiency is 0.2 the useful power is N = 1 MW. For other wind speed the useful power is: V = 5 m/s, N =0.58 MW; V = 6 m/s, N = 1 MW; V = 7 m/s, N =1.59 MW; V = 8 m/s, N=2.37 MW; V = 9 m/s, N =3.375 MW; V = 10 m/s, N = 4.63 MW.

Estimation of economic efficiency.

Let us take the cost of the installation $0.5 million, a useful life of 10 years and maintenance of $20,000/year. The energy produced in one year (when the wind has standard speed 6 m/s) is E = 1000x24x360 = 8.64 million kWh. The basic cost of energy is 70,000/8,640,000 = 0.81 cent/kWh.

Some technical parameters.

If the thrust is 23 tons, the tensile stress is 200 kg/sq.mm (composed fiber), then the parachute cable diameter is 12 mm, The full weight of the installation is 4.5 tons. The support wing has size 25x4 m.

Project 4. High speed air Darreus rotor at an altitude 1 km (fig.5).

Let us consider a rotor having the diameter of 100 m, a length of 200 m (work area is 20,000 sq.m). When the wind speed at an altitude H=10 m is V = 6 m/s, then at an altitude H = 1000 m it is 13 m/s. The full wind power is 13,46 MW. Let us take the efficiency coefficient 0.35, then the power of the Installation will be N = 4.7 MW. The change of power from wind speed is: V = 5 m/s, N = 2.73 MW; V = 6 m/s, N = 4.7 MW; V = 7 m/s, N = 7.5 MW; V = 8 m/s, N = 11.4 MW; V = 9m/s, N = 15.9 MW; V = 10 m/s, N = 21.8 MW. At an altitude of H = 13 km with an air density 0.267 and wind speed V = 40 m/s, the given installation will produce power N = 300 MW.

Estimation of economic feasibility.

Let us take the cost of the Installation at $1 million, a useful life of 10 years, and maintenance of $50,000/year. Our installation will produce E = 41 million kWh per year (when the wind speed equals

6 m/s at an altitude 10 m). The prime cost will be 150,000/41,000,000 = 0.37 cent/kWh. If the customer price is $0.15/kWh and profit from 1 kWh is $0.10 /kWh the profit will be $4.1 million per year.

Estimation of technical parameters.

The blade speed is 78 m/s. Numbers of blade is 4. Number of revolution is 0.25 revolutions per second. The size of blade is 200×0.67 m. The weight of 1 blade is 1.34 tons. The total weight of the Installation is about 8 tons. The internal wing has size 200×2.3 m. The additional wing has size 200x14.5 m and weight 870 kg. The cross-section area of the cable transmission having an altitude of $H = 1$ km is 300 sq.mm, the weight is 1350 kg.

Conclusion

Relatively no progress has been made in windmill technology in the last years. While the energy from wind is free, its production is more expensive than its production in conventional electric power stations. Conventional windmills are approached their maximum energy extraction potential relative to their installation cost. At present time the largest wind installations involves a tower with height up to 100 m, propeller diameter up to 154 m and power up to 5.6MW for wind speed 10 m/s. Current wind installations cannot essential decrease a cost of kWh, stability of energy production. They cannot continue increasing of power of single energy unit.

The renewable energy industry needs revolutionary ideas that improve performance parameters (installation cost and power per unit) and that significantly decreases (in 5-10 times) the cost of energy production. The airborne wind installations delineated in this paper can move the wind energy industry from stagnation to revolutionary potential.

The following is a list of benefits provided by the proposed high altitude new airborne wind systems compared to current grown installations:

1. The produced energy is least in 10 times cheaper than energy produced in conventional electric stations which includes current wind installation.
2. The proposed system is relatively inexpensive (no expensive tower), it can be made with a very large blades thus capturing wind energy from an enormous area (tens of times more than typical wind turbines).
3. The proposed installation does not require large ground space.
4. The installation may be located near customers and not require expensive high voltage equipment. It is not necessary to have long, expensive, high-voltage transmission lines and substations. Ocean going vessels can use this installation for its primary propulsion source.
5. Neither noise nor marring the landscape ruining the views.
6. The energy production is more stable because the wind is steadier at high altitude. The wind may be zero near the surface but it is typically strong and steady at higher altitudes. This can be observed when it is calm on the ground, but clouds are moving in the sky. There are a strong permanent air streams at a high altitude at many regions of the USA and World.
7. The installation can be easy relocated to other places.

As with any new idea, the suggested concept is in need of research and development. The theoretical problems do not require fundamental breakthroughs. It is necessary to design small, free flying installations to study and get an experience in the design, launch, stability, and the cable energy transmission from a flying wind turbine to a ground electric generator.

This paper has suggested some design solutions from patent application [2]. The author has many detailed analysis in addition to these presented projects. Organizations interested in these projects can address the author (http://Bolonkin.narod.ru , aBolonkin@juno.com , abolonkin@gmail.com).

The other ideas are in [1]-[6]. For example in [18-19] it is offered the electronic method is getting energy from air and water flow without turbines.

Acknowledgement

The author wishes to acknowledge Shmuel Neumann for correcting the English and offering useful advices and suggestions.

References

(Reader can find part of these articles in WEBs: http://Bolonkin.narod.ru/p65.htm, http://www.scribd.com(23); http://arxiv.org , (45); http://www.archive.org/ (20) and http://aiaa.org (41) search "Bolonkin").

1. Bolonkin A.A., Utilization of Wind Energy at High Altitude, AIAA-2004-5756, AIAA-2004-5705. International Energy Conversion Engineering Conference at Providence, RI, USA, Aug.16-19, 2004.
2. Bolonkin, A.A., "Method of Utilization a Flow Energy and Power Installation for It", USA patent application 09/946,497 of 09/06/2001.
3. Bolonkin, A.A., Transmission Mechanical Energy to Long Distance. AIAA-2004-5660.
4. Bolonkin, A.A., "New Concepts, Ideas, Innovations in Aerospace, Technology and the Human Sciences", NOVA, 2006, 510 pgs. http://www.scribd.com/doc/24057071 , http://www.archive.org/details/NewConceptsIfeasAndInnovationsInAerospaceTechnologyAndHumanSciences;
5. Bolonkin, A.A., "New Technologies and Revolutionary Projects", Lulu, 2008, 324 pgs, http://www.scribd.com/doc/32744477 , http://www.archive.org/details/NewTechnologiesAndRevolutionaryProjects,
6. Bolonkin, A.A., Cathcart R.B., "Macro-Projects: Environments and Technologies", NOVA, 2007, 536 pgs. http://www.scribd.com/doc/24057930 . http://www.archive.org/details/Macro-projectsEnvironmentsAndTechnologies
7. Gipe P., Wind Power, Chelsea Green Publishing Co., Vermont, 1998.
8. Thresher R.W. and etc, Wind Technology Development: Large and Small Turbines, NRFL, 1999.
9. Galasso F.S., Advanced Fibers and Composite, Gordon and Branch Scientific Publisher, 1989.
10. Carbon and High Performance Fibers Directory and Data Book, London-New York: Chapmen& Hall, 1995, 6th ed., 385 p.
11. Concise Encyclopedia of Polymer Science and Engineering, Ed. J.I.Kroschwitz, N.Y.,Wiley,1990,1341p.
12. Dresselhaus, M.S., Carbon Nanotubes, by, Springer, 2000.
13. Joby turbines. http://www.jobyenergy.com/tech.
14. Makani turbine: http://theenergycollective.com/energynow/69484/airborne-wind-turbine-could-revolutionize-wind-power , http://www.treehugger.com/wind-technology/future-wind-power-9-cool-innovations.html .
15. Cost of renewable energy. http://www.irena.org/DocumentDownloads/Publications/RE_Technologies_Cost_Analysis-WIND_POWER.pdf
16. Koshkin P., Shirkevuch M., Directory of Elementary Physics., Moscow, Nauka, 1982 (in Russian).
17. Wikipedia. Wind Energy.
Futher reading:
18. Bolonkin A.A., Electronic Wind Generator. Electrical and Power Engineering Frontier Sep. 2013, Vol. 2 Iss. 3, pp. 64-71. http://www.academicpub.org/epef/Issue.aspx?Volume=2&Number=3&Abstr=false , http://viXra.org/abs/1306.0046
19. Bolonkin AA., Electron Hydro Electric Generator. International Journal of Advanced Engineering Applications. ISSN: 2321-7723 (Online), Special Issue I, 2013. http://fragrancejournals.com/?page_id=18, http://viXra.org/abs/1306.0196,

24 April 2013

Chapter 3

Electrostatic Generator and Electronic Transformer

Abstract

Transmission of high voltage by direct currency (DC) over long distance has big advantages in comparison with the current transmission by alternative current (AC). But about one hundred years ago generating electricity by the magnetic method was easy and development led to a system centered on AC. Now AC is the dominant method. That is only a result of the quirks of electricity's historical development.

In present time it is possible to research and develop (R&D) high voltage electrostatic generators and inverters of high voltage DC in low voltage DC and in AC and using the old AC lines and devices.

Author offers a cheap high voltage electrostatic generator and transformer of high voltage DC to low voltage DC and DC in AC (any frequency and phases) and back. That may be adopted gradually and it will give significantly savings of electricity and energy.

Key words: Electrostatic generator. Electronic transformer. High voltage generator, long distance electric line.

Introduction

Electric Generator.

In electricity generation, an electric generator is a device that converts mechanical energy to electrical energy. A generator forces electric current to flow through an external circuit. The source of mechanical energy may be a reciprocating or turbine steam engine, water falling through a turbine or waterwheel, an internal combustion engine, a wind turbine, a hand crank, compressed air, or any other source of mechanical energy. Generators provide nearly all of the power for electric power grids.

The reverse conversion of electrical energy into mechanical energy is done by an electric motor, and motors and generators have many similarities. Many motors can be mechanically driven to generate electricity and frequently make acceptable generators.

Electrostatic generator,

is a mechanical device that produces *static electricity*, or electricity at high voltage and low continuous current. The knowledge of static electricity dates back to the earliest civilizations, but for millennia it remained merely an interesting and mystifying phenomenon, without a theory to explain its behavior and often confused with magnetism. By the end of the 17th Century, researchers had developed practical means of generating electricity by friction, but the development of electrostatic machines did not begin in earnest until the 18th century, when they became fundamental instruments in the studies about the new science of electricity. Electrostatic generators operate by using manual (or other) power to transform mechanical work into electric energy. Electrostatic generators develop electrostatic charges of opposite signs rendered to two conductors, using only electric forces, and work by using moving plates, drums, or belts to carry electric charge to a high potential electrode. The charge is generated by one of two methods: either the triboelectric effect (friction) or electrostatic induction.

Electric-power transmission

is the bulk transfer of electrical energy, from generating power plants to electrical substations located near demand centers. This is distinct from the local wiring between high-voltage substations and customers, which is typically referred to as electric power distribution. Transmission lines, when interconnected with each other, become transmission networks. The combined transmission and distribution network is known as the "power grid" in the United States, or just "the grid". In the United Kingdom, the network is known as the "National Grid".

A wide area synchronous grid, also known as an "interconnection" in North America, directly connects a large number of generators delivering AC power with the same relative phase, to a large number of consumers. For example, there are three major interconnections in North America (the Western Interconnection, the Eastern Interconnection and the Electric Reliability Council of Texas (ERCOT) grid), and one large grid for most of continental Europe.

The electric transmission may be overhead and underground. High-voltage overhead conductors are not covered by insulation. The conductor material is nearly always an aluminum alloy, made into several strands and possibly reinforced with steel strands.

Today, transmission-level voltages are usually considered to be 110 kV and above. Lower voltages such as 66 kV and 33 kV are usually considered subtransmission voltages but are occasionally used on long lines with light loads. Voltages less than 33 kV are usually used for distribution. Voltages above 230 kV are considered extra high voltage and require different designs compared to equipment used at lower voltages.

Electric power can also be transmitted by underground power cables instead of overhead power lines. Underground cables take up less right-of-way than overhead lines, have lower visibility, and are less affected by bad weather. However, costs of insulated cable and excavation are much higher than overhead construction.

Engineers design transmission networks to transport the energy as efficiently as feasible, while at the same time taking into account economic factors, network safety and redundancy. These networks use components such as power lines, cables, circuit breakers, switches and transformers.

Transmission and distribution losses in the USA were estimated at 6.6% in 1997 and 6.5% in 2007.

As of 1980, the longest cost-effective distance for direct-current transmission was determined to be 7,000 km (4,300 mi). For alternating current it was 4,000 km (2,500 mi), though all transmission lines in use today are substantially shorter than this.

In any alternating current transmission line, the inductance and capacitance of the conductors can be significant. Currents that flow solely in 'reaction' to these properties of the circuit, (which together with the resistance define the impedance) constitute reactive power flow, which transmits no 'real' power to the load.

These reactive currents however are **very** real and cause extra heating losses in the transmission circuit. The ratio of 'real' power (transmitted to the load) to 'apparent' power (sum of 'real' and 'reactive') is the power factor. As reactive current increases, the reactive power increases and the power factor decreases. For transmission systems with low power factor, losses are higher than for systems with high power factor. Utilities add capacitor banks, reactors and other components (such as phase-shifting transformers; static VAR compensators; physical transposition of the phase conductors; and flexible AC transmission systems, FACTS) throughout the system to compensate for the reactive power flow and reduce the losses in power transmission and stabilize system voltages. These measures are collectively called 'reactive support.

High-voltage direct current.

High-voltage direct current (HVDC) is used to transmit large amounts of power over long distances or for interconnections between asynchronous grids. When electrical energy is to be transmitted over very long distances, the power lost in AC transmission becomes appreciable and it is less expensive to use direct current instead of alternating current. For a very long transmission line, these lower losses (and reduced construction cost of a DC line) can offset the additional cost of the required converter stations at each end.

HVDC is also used for submarine cables because over about 30 kilometres (19 mi) lengths AC cannot be supplied. In these cases special high voltage cables for DC are used. Submarine HVDC systems are often used to connect the electricity grids of islands, for example, between Great Britain and mainland Europe, between Great Britain and Ireland, between Tasmania and the Australian mainland, and between the North and South Islands of New Zealand. Submarine connections up to 600 kilometres (370 mi) in length are presently in use.

HVDC links can be used to control problems in the grid with AC electricity flow. The power transmitted by an AC line increases as the phase angle between source end voltage and destination ends increases, but too large a phase angle will allow the systems at either end of the line to fall out of step. Since the power flow in a DC link is controlled independently of the phases of the AC networks at either end of the link, this phase angle limit does not exist, and a DC link is always able to transfer its full rated power. A DC link therefore stabilizes the AC grid at either end, since power flow and phase angle can then be controlled independently.

As an example, to adjust the flow of AC power on a hypothetical line between Seattle and Boston would require adjustment of the relative phase of the two regional electrical grids. This is an everyday occurrence in AC systems, but one that can become disrupted when AC system components fail and place unexpected loads on the remaining working grid system. With an HVDC line instead, such an interconnection would: (1) Convert AC in Seattle into HVDC. (2) Use HVDC for the three thousand miles of cross-country transmission. Then (3) convert the HVDC to locally synchronized AC in Boston, (and possibly in other cooperating cities along the transmission route). Such a system could be less prone to failure if parts of it were suddenly shut down. One example of a long DC transmission line is the Pacific DC Intertie located in the Western United States.

The cost of high voltage electricity transmission (as opposed to the costs of electric power distribution) is comparatively low, compared to all other costs arising in a consumer's electricity bill. In the UK transmission costs are about 0.2p/kWh compared to a delivered domestic price of around 10 p/kWh.

Research evaluates the level of capital expenditure in the electric power T&D equipment market will be worth $128.9bn in 2011.

Advantages of HVDC over AC transmission.
The most common reason for choosing HVDC over AC transmission is that HVDC is more economic than AC for transmitting large amounts of power point-to-point over long distances. A long distance, high power HVDC transmission scheme generally has lower capital costs and lower losses than an AC transmission link.

Even though HVDC conversion equipment at the terminal stations is costly, overall savings in capital cost may arise because of significantly reduced transmission line costs over long distance routes. HVDC needs fewer conductors than an AC line, as there is no need to support three phases. Also, larger conductors can be used since HVDC does not suffer from the skin effect. These factors can lead to large reductions in transmission line cost for a long distance HVDC scheme.

Depending on voltage level and construction details, HVDC transmission losses are quoted as about 3.5% per 1,000 km, which is less than typical losses in an AC transmission system. HVDC transmission may also be selected because of other technical benefits that it provides for the power system. HVDC schemes can transfer power between separate AC networks. HVDC powerflow between separate AC systems can be automatically controlled to provide support for either network during transient conditions, but without the risk that a major power system collapse in one network will lead to a collapse in the second.

The combined economic and technical benefits of HVDC transmission can make it a suitable choice for connecting energy sources that are located far away from the main load centers.

Specific applications where HVDC transmission technology provides benefits include:

Undersea cables transmission schemes (e.g., 250 km Baltic Cable between Sweden and Germany, the 580 km NorNed cable between Norway and the Netherlands, and 290 km Basslink between the Australian mainland and Tasmania).

Endpoint-to-endpoint long-haul bulk power transmission without intermediate 'taps', usually to connect a remote generating plant to the main grid, for example the Nelson River DC Transmission System in Canada.

Increasing the capacity of an existing power grid in situations where additional wires are difficult or expensive to install.

Power transmission and stabilization between unsynchronised AC networks, with the extreme example being an ability to transfer power between countries that use AC at different frequencies. Since such transfer can occur in either direction, it increases the stability of both networks by allowing them to draw on each other in emergencies and failures.

Stabilizing a predominantly AC power-grid, without increasing fault levels (prospective short circuit current).

HVDC Converter.
At the heart of an HVDC converter station, the equipment which performs the conversion between AC and DC is referred to as the *converter*. Almost all HVDC converters are inherently capable of converting from AC to DC (*rectification*) and from DC to AC (*inversion*), although in many HVDC systems, the system as a whole is optimised for power flow in only one direction. Irrespective of how the converter itself is designed, the station which is operating (at a given time) with power flow from AC to DC is referred to as the *rectifier* and the station which is operating with power flow from DC to AC is referred to as the *inverter*.

Early HVDC systems used electromechanical conversion (the Thury system) but all HVDC systems built since the 1940s have used electronic (static) converters. Electronic converters for HVDC are divided into two main categories: Line-commutated converters (LCC), Voltage-sourced converters, or current-source converters.

A magnetic transformer
is an electrical device that transfers energy between two or more circuits through electromagnetic induction. A varying current in the transformer's primary winding creates a varying magnetic flux in the core and a varying magnetic field impinging on the secondary winding. This varying magnetic field at the secondary induces a varying electromotive force (emf) or voltage in the secondary winding. Making use of Faraday's Law in conjunction with high magnetic permeability core properties, transformers can thus be designed to efficiently change AC voltages from one voltage level to another within power networks.

Transformers range in size from RF transformers a small cm³ fraction in volume to units interconnecting the power grid weighing hundreds of tons. A wide range of transformer designs are used in electronic and electric power applications. Since the invention in 1885 of the first constant potential transformer, transformers have become essential for the AC transmission, distribution, and utilization of electrical energy.

Costs of high voltage DC transmission.
Generally, providers of HVDC systems, such as Alstom, Siemens and ABB, do not specify cost details of particular projects. It may be considered a commercial matter between the provider and the client.
Costs vary widely depending on the specifics of the project (such as power rating, circuit length, overhead vs. cabled route, land costs, and AC network improvements required at either terminal). A detailed comparison of DC vs. AC transmission costs may be required in situations where there is no clear technical advantage to DC alone, and economical reasoning drives the selection.
However, some practitioners have provided some information:
For an 8 GW 40 km link laid under the English Channel, the following are approximate primary equipment costs for a 2000 MW 500 kV bipolar conventional HVDC link (exclude way-leaving, on-shore reinforcement works, consenting, engineering, insurance, etc.)

Converter stations ~£110M (~€120M or $173.7M)

Subsea cable + installation ~£1M/km (~€1.2M or $6M/km)

So for an 8 GW capacity between England and France in four links, little is left over from £750 M for the installed works. Add another £200–300M for the other works depending on additional onshore works required.
An April 2010 announcement for a 2,000 MW, 64 km line between Spain and France is estimated at €700 million. This includes the cost of a tunnel through the Pyrenees.

Electric generator

Current electric grid needs a great deal of electricity which can be gotten either from a nuclear reactor, steam turbine or from connection of the conventional turbojet engine with an electric generator (gas turbine). Let us consider the last possibility.
When industry needs electricity, most electric engineers offer the conventional way: take the usual magnetic electric generator and connect it to the turbojet or other (for example, piston) engine.

Let us analyze the limiting possibilities of different versions.

Magnetic electric generator.
Magnetic electric generator was first produced about century ago and has been very well studied. The ratio of power/mass of magmatic generator for 1 m³ may be estimated by equation:

$$A = \frac{B^2}{2\mu_0}, \quad P = Av, \quad M = \gamma, \quad \frac{P}{M} = c\frac{B^2 v}{2\mu_0 \gamma}, \tag{1}$$

Here A is density of energy into 1 m³ of magnetic material J/m³; B is maximal magnetic intensity, T; $\mu_0 =$

$4\pi 10^{-7}$ is magnetic permeability (magnetic constant), N/A²; P is power, W; v is electric frequency, 1/s ($v = 50 \div 400$ 1/s, maximal $v \approx 700$ 1/s); M is mass 1 m³ of generator, kg/m³; γ is specific mass of the generator bogy, kg/m³ ($\gamma \approx 8000$ kg/m³); $c \approx 1/8$ correction coefficient, because average $B = 0.5 B_{max}$ and ferromagnetic iron uses only about ½ engine volume. The maximal frequency determinates the ratio L/r, where L is inductance, r is electric resistance. That equals about 500 – 1000 1/s.

Example. Let us take the typical data $B = 1$ T, $v = 400$ 1/s, $\gamma = 8000$ kg/m³. We get maximal $P/M = 2.5$ kW/kg.

Typical high performance aviation generator has:

Type: ГТ-120ПЧ8 (Russian)	Power	Phases	Voltage	Currency	Frequency	Number of rev.	Mass
	120 kW	3	208V	334 A	400 1/s	8000 in min	90 kg

The ratio for the usual aircraft generator equals 1.33 kW/kg. That is two times less than maximal possible. For our purposes that will be two times less because we need high voltage. But the high voltage transformer will weigh not less than electric generator. If aircraft has turbo 10,000 kW the magnetic propulsion system will weigh about 14 tons, 5 times more than turbojet. That is not acceptable in aviation and bad for a ground station. In addition, magnetic generator produces AC. The DC electrostatic generator weighs significantly less.

Electrostatic generator (EG).

Principal schema of electrostatic generator is in fig. 1.

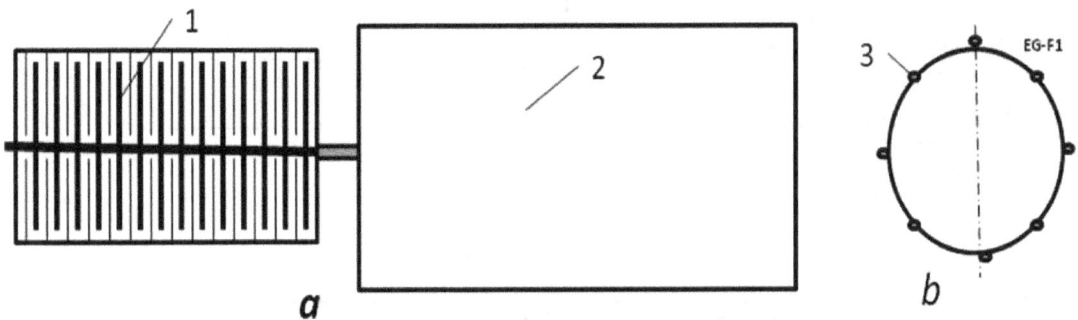

Fig. 1. Electrostatic generator. Notations: *a* – side view; *b* – forward view; 1 – electrostatic generator, 2 – steam, turbojet or other engine, 3 – electric collector.

The installation contains the electrostatic generator 1, the engine 2 (steam, turbo or other engine), electric collector 3 for charging the generator and getting the electricity.

The offer electrostatic generator is working the following way. There are stationary and rotate charged plates of capacitor (or two plates rotate in opposed direction). When charged plates remove from one to other, the voltage between them increases. When voltage reaches a need value, the plates discharges throw an outer load, After this they charge again and process is repeated.

Different types of electrostatic electric generator is known for about two centuries but it is not used because it produces very high voltage which is very dangerous for people and not suitable for practice and home devises. As a result, EG is studied very little and no power EG produces by industry.

The ratio power/mass of electrostatic generator for area $S = 1$ m^2 may be estimated by equation:

$$A = \frac{CU^2}{2}, \quad C = \frac{\varepsilon\varepsilon_0 S}{d}, \quad U = Ed, \quad P = \frac{A}{t}, \quad M = \gamma\delta S, \quad \frac{P}{M} = \frac{\varepsilon\varepsilon_0 E^2 \delta}{2\gamma d^2}V_a, \tag{2}$$

where A is density of energy on 1 m^2 of the electrostatic (isolator plate) material J/m^2; C is capacitance of plate (one plate of condenser), F/m^2; U is voltage, V; ε (1 ÷ 3000) dielectric constant of plate matter; $\varepsilon_0 = 8.85 \cdot 10^{-12}$ is permittivity, F/m; S is area of one plate, m^2; d is distance between plates (include thickness of one plate, m); P is power, W; t is time, s; M is mass 1 m^2 of generator plate, kg/m^3; γ is specific mass of the generator plate, kg/m^3 ($\gamma \approx 1800$ kg/m^3), δ is clearance between plates, m; V_a is the average relative speed of two plates, m/s ($V_a \approx 0.5V$, where V is the peripheral disk (plate) speed).

Properties of some insulators in Table 1.

Table 1. Properties of various good insulators (recalculated in metric system)

Insulator	Resistivity Ohm-m.	Dielectric strength MV/m.. E_i	Dielectric constant, ε	Tensile strength kg/mm^2, $\sigma \times 10^7$ N/m^2
Lexan	10^{17}–10^{19}	320–640	3	5.5
Kapton H	10^{19}–10^{20}	120–320	3	15.2
Kel-F	10^{17}–10^{19}	80–240	2–3	3.45
Mylar	10^{15}–10^{16}	160–640	3	13.8
Parylene	10^{17}–10^{20}	240–400	2–3	6.9
Polyethylene	10^{18}–5×10^{18}	40–680*	2	2.8–4.1
Poly (tetra-fluoraethylene	10^{15}–5×10^{19}	40–280**	2	2.8–3.5
Air (1 atm, 1 mm gap)	-	4	1	0
Vacuum (1.3×10^{-3} Pa, 1 mm gap)	-	80–120	1	0

*For room temperature 500–700 MV/m.
** 400–500 MV/m.

Source: Encyclopedia of Science & Technology[9] (Vol. 6, p. 104, p. 229, p. 231).

Note: Dielectric constant ε can reach 4.5 – 7.5 for mica (E is up 200 MV/m); 6 – 10 for glasses ($E = 40$ MV/m) and 900 – 3000 for special ceramics (marks are CM-1, T-900) ($E = 13 – 28$ MV/m). Dielectric strength appreciable depends from surface roughness, thickness, purity, temperature and other conditions of material. It is necessary to find good insulating materials and reach conditions which increase the dielectric strength.

The safe peripheral disk speed may be estimated by equation $V = (\sigma/\gamma)^{0.5}$ where σ is safety tensile stress (N/m^2), γ is specific weight, kg/m^3. The disk may be reinforced by fiber having high tensile stress.

Let us consider the following experimental and industrial fibers, whiskers, and nanotubes:

Experimental nanotubes CNT (carbon nanotubes) have a tensile strength of 200 Giga-Pascals (20,000 kg/mm^2), Young's modulus is over 1 Tera Pascal, specific density γ=1800 kg/m^3 (1.8 g/cc) (year

2000). For safety factor $n = 2.4$, $\sigma = 8300$ kg/mm^2 = 8.3×10^{10} N/m^2, $\gamma = 1800$ kg/m^3, $(\sigma/\gamma) = 46 \times 10^6$, $K = 4.6$. The SWNTs nanotubes have a density of 0.8 g/cc, and MWNTs have a density of 1.8 g/cc. Unfortunately, the nanotubes are very expensive at the present time (1994).

For whiskers $C_D \sigma = 8000$ kg/mm^2, $\gamma = 3500$ kg/m^3 (1989).

For industrial fibers $\sigma = 500 - 600$ kg/mm^2, $\gamma = 1800$ kg/m^3, $\sigma/\gamma = 2{,}78 \times 10^6$, $K = 0.278 - 0.333$,

Figures for some other experimental whiskers and industrial fibers are given in Table 2.

Table 2. Properties of fiber and whiskers

Material Whiskers	Tensile strength kgf/mm^2	Density g/cc	Material Fibers	Tensile strength MPa	Density g/cc
AlB$_{12}$	2650	2.6	QC-8805	6200	1.95
B	2500	2.3	TM9	6000	1.79
B$_4$C	2800	2.5	Thorael	5650	1.81
TiB$_2$	3370	4.5	Allien 1	5800	1.56
SiC	1380–4140	3.22	Allien 2	3000	0.97

See Reference [9] p. 33.

Example: Let us estimate ratio P/M of the electrostatic generator by equation (2). Take the electric intensity $E = 10^7$ V/m, area of the disk 1 m^2, thickness of the disk 0.003m, clearance between disks $\delta = 0{,}002$ m, ($d = 0.005$ m), $V = 500$ m/s, $\gamma = 1800$ kg/m^3, $\varepsilon = 3$. Substitute these data in equation (23) we get $P/M = 53$ kW/kg. That means the electrostatic generator (motor) of equal power will be in 20 times less than magnetic generator (motor). The 10,000 kW electrostatic generator (motor) will weigh only 400 kg (200 disks). And additional the electrostatic generator produces high voltage direct (constant) electric currency (DC).

Air friction in electrostatic generator and its efficiency.

Let us estimate ratio of the air friction/produced power 1 m^2 of disk the electrostatic generator. Compute the friction, produced power and efficiency:

$$P_f = 2FV_a, \quad F = \varsigma \frac{V_a}{\delta}, \quad P_f = 2\frac{\varsigma V_a^2}{\delta}, \quad P = \frac{\varepsilon \varepsilon_0 E^2}{2} V_a, \quad \eta = 1 - \frac{P_f}{P} = 1 - \frac{4\varsigma V_a}{\varepsilon \varepsilon_0 \delta E^2}, \quad (3)$$

where P_f is power of friction 1 m^2 of disk, W/m^2; F is friction force 1 m^2 of disk, N/m^2; V_a is average disk speed, m/s; ς is viscosity of the gas (for air $\varsigma = 1.72 \cdot 10^{-5}$ Pa·s, for hydrogen $\varsigma = 0.84 \cdot 10^{-5}$ at atmospheric pressure and $T = 0°$C); P is power produced 1 m^2 of disk, W/m^2; ε is dielectric constant of plate matter; $\varepsilon_0 = 8{,}85 \cdot 10^{-12}$ is electric permeability, F/m; δ is clearance between disk, m; E is electric intensity, V/m; η is efficiency of generator related to air friction.

Example: If $V_a = 250$ m/s; $E = 2 \cdot 10^6$ V/m; $\delta = 0.002$ m, then $\eta = 0.92$.

The coefficient of gas friction weak depends from the pressure and temperature. If we change the air into the electrostatic generator by hydrogen, the loss of friction decreases in two times. If we create a vacuum in the electrostatic generator, the gas friction will be zero and the safety electric intensity is increased in many times.

Transformer direct currency into alternative currency and back.

Principal schema of the offered electronic transformer - high voltage DC in low voltage DC or DC in AC and back is presented in fig.2.

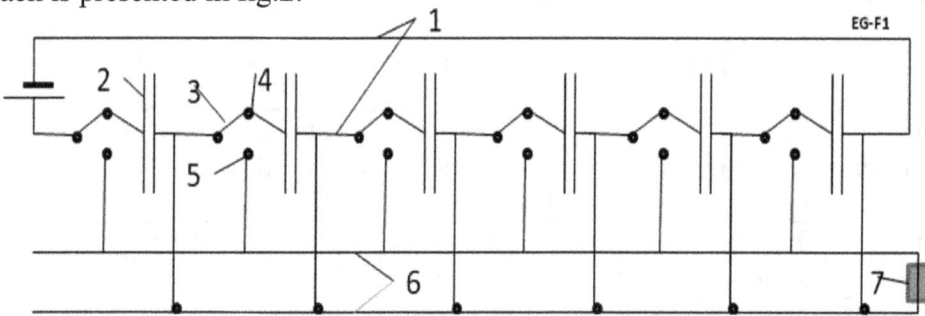

Fig.2. Principal Schema of the offered transformer -- high voltage DC in low voltage DC or DC in AC and back. *Notations*: 1 – high voltage line, 2 – condensers (capacitor), 3 – two positions electronic switch, 4 – contact the high voltage line, 5 – contact the low voltage line, 6 – low voltage line. 7 - outer electric load.

The installation contains two line (high voltage 1 and low voltage 6), the series of the capacitors 2 and series of the electronic switches 3. Switches 3 have a two position 4, 5. In position 4 the switches connect the capacitors in the series connection; in position 5 the switches connect the capacitors in the parallel connection. In position 4 every capacitor has voltage U/n, where U is voltage of line 1, n is number of capacitors in series. In position 5 the voltage is U/n in line 6, but an electric currency increases in n times. As the result we decrease voltage U in n times, but increases the currency i in n times.

The offered transformer can work in back direction: increases the voltage of DC to very high voltage. Connections the capacitor and electronic switches we can transform the direct currency in the any alternative currency, any frequency, any phases, any form of waves (directly and back).

The possible frequency determined by ratio $1/rC$, where C is capacitance of capacitor, r is Ohm electrical resistance. That equals about $10^8 – 10^9$ 1/s.

SUMMARY.

The author offers a cheap electrostatic generator produced a high voltage direct currency (DC) and electronic transformer, which converts high voltage DC into low voltage DC or any alternative electric currency (or back). He shows a transmission of the high voltage direct currency (DC) to long distance has big advantages in comparison with the current transmission by the alternative current (AC). The electrostatic generator (installation) is cheaper, more efficiency and more appropriate for long distance transmission than current equipment for AC. one hundred years ago generating electricity by the magnetic method was easy and development led to a system centered on AC. Now AC is the dominant method. That is only a result of the quirks of electricity's historical development. But development of electronics in recent years makes possible the design of powerful electrostatic generators and electronic transformers.

In present time it is achievable to research and develop (R&D) the high voltage electrostatic generators and inverters of high voltage DC in low voltage DC and in AC and using the old AC lines and devices.

In present time it is possible to research and develop (R&D) high voltage electrostatic generators and inverters of high voltage DC in low voltage DC and in AC and using the old AC lines and devices.

Author offers a cheap high voltage electrostatic generator and transformer of high voltage DC to low voltage DC and DC in AC (any frequency and phases) and back. That may be adopted gradually and it will give significantly savings of electricity and energy.

Researches related to this topic are presented in [1]-[13].

ACKNOWLEDGEMENT

The author wishes to acknowledge Joseph Joseph Friedlander for correcting the English and offering useful advice and suggestions.

References

[1]. A.A. Bolonkin, *"New Concepts, Ideas, Innovations in Aerospace, Technology and the Human Sciences"*, NOVA, 2006, 510 pgs. **ISBN-13: 978-1-60021-787-6.**
http://www.archive.org/details/NewConceptsIfeasAndInnovationsInAerospaceTechnologyAndHumanSciences, http://www.scribd.com/doc/24057071 ,

[2]. A.A. Bolonkin, R. Cathcart, *"Macro-Projects: Environments and Technologies"*, NOVA, 2007, 536 pgs. ISBN 978-1-60456-998-8. http://www.scribd.com/doc/24057930; http://www.archive.org/details/Macro-projectsEnvironmentsAndTechnologies .

[3]. A.A. Bolonkin, *Femtotechnologies and Revolutionary Projects*. Lambert, USA, 2011. 538 p., 16 Mb.
http://www.scribd.com/doc/75519828/ ,
http://www.archive.org/details/FemtotechnologiesAndRevolutionaryProjects

[4] A.A. Bolonkin , *Innovations and New Technologies*. Scribd, 30/7/2013. 309 pgs. 8 Mb.
http://www.scribd.com/doc/157098739/Innovations-and-New-Technologies-7-9-13
http://archive.org/details/InnovationsAndNewTechnologies, http://viXra.org/abs/1307.0169

[5] A.A.Bolonkin, Electrostatic Climber for Space Elevator and Launcher. Paper AIAA-2007-5838 for *43 Joint Propulsion Conference*. Cincinnati, Ohio, USA, 9 – 11 July, 2007. See also [12], Ch.4, pp. 65-82.

[6] A.A.Bolonkin, Wireless Transfer of Electricity from Continent to Continent.
International Journal of Sustainable Engineering. 2011. Vol.4, #4, p. 290-300.
http://www.scribd.com/doc/42721638/,
http://www.archive.org/details/WirelessTransferOfElectricityFromContinentToContinent

[7] A.A.Bolonkin, Hypersonic Ground Electric AB Engine.
Academic Journal of Applied Sciences Research (AJASR), Volume-1, Issue–1, 2013.
http://www.scribd.com/doc/119462908/Hypersonic-Ground-Electric-AB-Engine
http://archive.org/details/HypersonicGroundElectricAbEngine
International Journal of Advanced Engineering Applications, Vol.1, Iss.4, pp.32-43 (2012)
http://fragrancejournals.com/wp-content/uploads/2013/03/IJAEA-1-4-4.pdf

[8] A.A. Bolonkin, Air Catapult Transportation. NY, USA, Scribd, 2011.
Journal of Intelligent Transportation and Urban Planning (JTUP), April 2014, Vol.2, pp. 70-84.
http://www.scribd.com/doc/79396121/Article-Air-Catapult-Transportation-for-Scribd-1-25-12,
http://www.archive.org/details/AirCatapultTransport, http://viXra.org/abs/1310.0065 .
Chapter in Book: Recent Patents on Electrical & Electronic Engineering, Bentham Science Publishers, Vol.5, No.3, 2012.

[9] A.A. Bolonkin, *Non Rocket Space Launch and Flight*. Elsevier, 2005. 488 pgs. ISBN-13: 978-0-08044-731-5, ISBN-10: 0-080-44731-7 .
http://www.scribd.com/doc/203941769/Non-Rocket-Space-Launch-and-Flight-v-3 ,
http://www.archive.org/details/Non-rocketSpaceLaunchAndFlight ,
http://www.scribd.com/doc/24056182 .

[10] S.G. Kalashnikov, Electricity, Moscow, Nauka, 1985.(in Russian).

[11] N.I. Koshkin and M.G. Shirkebich, Directory of Elementary Physics, Nauka, Moscow, 1982 (in Russian).

[12] I.K. Kikoin. Table of Physics values. Atomisdat, Moscow, 1976 (in Russian).
[13] Wikipedia. Electricity, http://wikipedia.org .

June 30, 2014

Chapter 4

Jet Electric Generator

Abstract

Author offers and develops the theory of a new simple cheap efficient electric (electron) generator. This generator can convert pressure or kinetic energy of any non-conductive flow (gas, liquid) into direct current (DC). The generator can convert the mechanical energy of any engine into high voltage DC. One can covert the wind and water energy into electricity without turbine. One can convert the rest energy of an internal combustion engine or turbojet engine in electricity and increase its efficiency.

Key words: Jet Electric Generator, Electron generator, AB generator, Wind electric generator, Water electric generator, DC generator, High voltage generator.

Introduction

Electric Generator.
In electricity generation, an electric generator is a device that converts mechanical energy to electrical energy. A generator forces electric current to flow through an external circuit. The source of mechanical energy may be a reciprocating or turbine steam engine, water falling through a turbine or waterwheel, an internal combustion engine, a wind turbine, a hand crank, compressed air, or any other source of mechanical energy. Generators provide nearly all of the power for electric power grids.

The reverse conversion of electrical energy into mechanical energy is done by an electric motor, and motors and generators have many similarities. Many motors can be mechanically driven to generate electricity and frequently make acceptable generators.

The **MHD** (magneto hydrodynamic) **generator** transforms thermal energy and kinetic energy directly into electricity. MHD generators are different from traditional electric generators in that they operate at high temperatures without moving parts. MHD was developed because the hot exhaust gas of an MHD generator can heat the boilers of a steam power plant, increasing overall efficiency. MHD was developed as a topping cycle to increase the efficiency of electric generation, especially when burning coal or natural gas. MHD dynamos are the complement of MHD propulsors, which have been applied to pump liquid metals and in several experimental ship engines.

An MHD generator, like a conventional generator, relies on moving a conductor through a magnetic field to generate electric current. The MHD generator uses hot conductive plasma as the moving conductor. The mechanical dynamo, in contrast, uses the motion of mechanical devices to accomplish this. MHD generators are technically practical for fossil fuels, but have been overtaken by other, less expensive technologies, such as combined cycles in which a gas turbine's or molten carbonate fuel cell's exhaust heats steam to power a steam turbine.

Natural MHD dynamos are an active area of research in plasma physics and are of great interest to the geophysics and astrophysics communities, since the magnetic fields of the earth and sun are produced by these natural dynamos.

The jet electric generator offered in given article is principal different from MHD generator. One does not need hot plasma, magnets and a magnetic field, it's easier, cheaper by ten times and more efficient.

It might serve for purposes of propulsion. MHD is also not reversible.

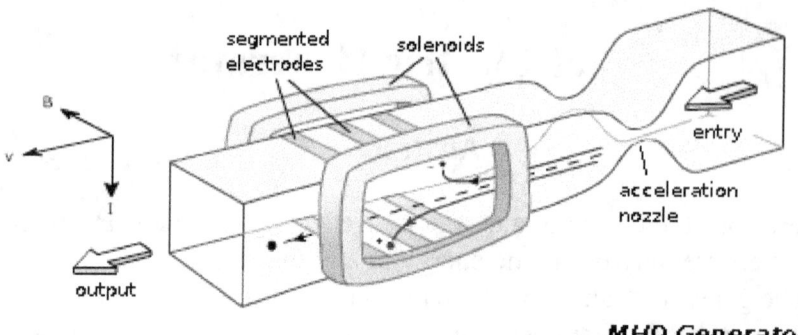

The first author publications about new jet AB electron-electric generator are in [1] – [5].

AB Jet Electric Generator (ABJEG)

Principal schema. Jet electric generator (ABJEG) is very simple (fig.2). That is nonconductive tube 2, injector 4 of electrons (ions) in beginning of tube and collector 5 of electrons (ions) in end of tube. If generator does not have grounding 10, one may have the charge compensator 10.

The electron injector is conventional: cold field electron emission (edge needles) or hot electron emission (hot cathode). See more detailed description and computation of the injectors in next chapters. Correct design of them practically does not consume electricity.

The charge (electron) collector may be conductive plates or conductive net located in end of tube. The ABJEG needs in it if one does not have the good grounding or want to improve the efficiency.

The charge compensator deletes the opposed charges (electrons and positive ions) and injects the surplus charges into exhaust flow.

The charge compensator is necessary if ABJEG cannot have the grounding (for example ABJEG is located in aircraft).

The offered generator can work on non-conductive gas or liquid. It may be convertible to either a pump or propulsion system.

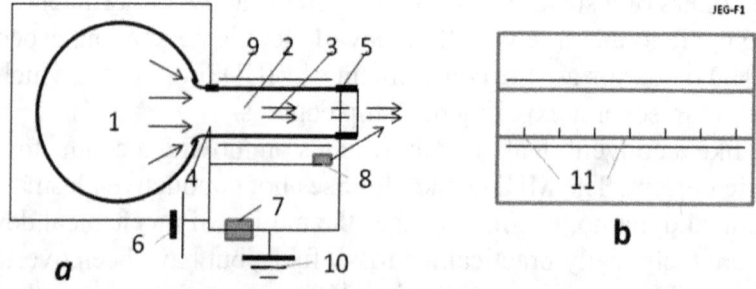

Fig.2. Schema of AB Jet electric generator. *a* – side view; *b* – back view. *Notations:* 1 is pressured gas or liquid, 2 is Jet Electric Generator (ABJEG), 3 is flow, 4 is injector of electrons, 5 is collector of electrons, 6 is source of injector voltage, 7 is useful load, 8 is compensator of the lost electrons, 9 is compensator of internal charge, 10 is grounding (if no grounding we need the compensator 8), 11 needles of the ejector 4.

Work of the ABJEG (fig. 3). The nonconductive pressure gas (or liquid) locates into volume 1 (fig.2). Under pressure the gas flow into ABJEG (fig.3, tube 1). In beginning of tube the injector 2 injects into gas the electrons 5. The electrons are captured by the flow 6 and move to end of tube 1 against the electric field between injector and collector. They brake the flow (get the work, create the opposed pressure). The flow reaches the collector (plate 7, 11) and charges it. When the charge of collector became over the charge of the injector 4 (fig.2) the electrical current appears in the circuit. It consume by load 7 (fig.2).

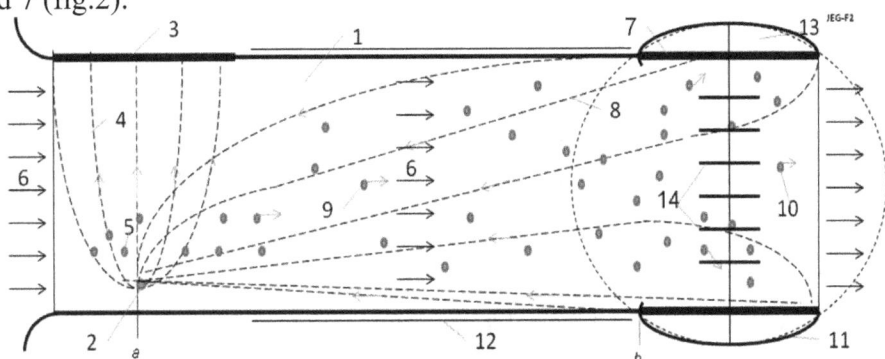

Fig.3. Work of the AB Jet Electric Generator. *Notations*: 1 is Jet Electric Generator, 2 is needle injector, 3 is opposed cathode plate of needle injector, 4 is the electric intensity lines of needle injector, 5 is the injected electrons, 6 is flow of working mass (gas or liquid), 7 is plate of collector, 8 is the electric intensity lines of the needle injector, 9 is electron moved against the electric field under flow pressure, 10 is electron not captured collector, 11 conductive isolated surface of collector, 12 is conductive coating of the dielectric tube 1 for balancing the internal charge of electrons, 13 is collector of electrons, 14 are conductive internal plates of collector.

Different designs of the injectors, collectors, compensators and electric schemas of ABJEG are possible.
One of them is shown in fig. 4. This injector has a conductivity net a high transparency and collector having opposed charged plates. This collector attracts the electron and increases the efficiency. Correct design of them practically does not consume electricity.

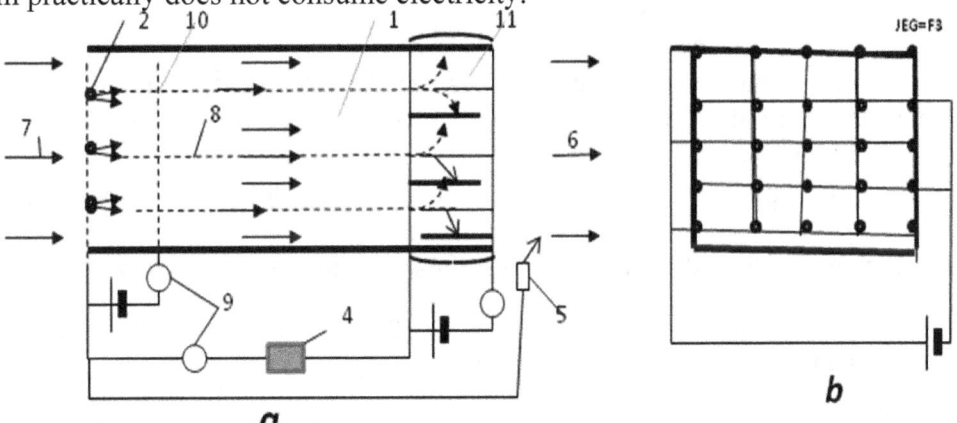

Fig.4. Electric schema of one version of the ABJEG. *Notations*: 2 is electron injector, 4 is useful load, 5 is compensator, 6 is exhaust flow, 7 is input stream, 8 is trajectory of electrons, 9 is control, 10 is anode net of injector, 11 is collector with opposed charged plates.

Differences of ABJEG (AB Electric Generator) relative to MHD (magnetohydrodynamic generator).
The jet electric generator is principal different from MHD generator. MHD works on plasma or conductive

liquid. ABJEG works on dielectric (non-conductive gas or liquid (for example, water)). MHD needs in very hot gas (high conductivity plasma). The currently available materials cannot endure this temperature. Result in practical systems is low efficiency. Converting of gas to a high conductivity plasma requests a lot of energy for heating, ionization and dissociation of gas. Most part of this energy is useless loses. The MHD needs very powerful magnets (better superconductive magnets). For increasing efficiency the MHD connects to conventional gas turbine. The installation is very complex and expensive. MHD is not reversible.

Advantages of ABJEG over MHD:

1. ABJEG does not need hot plasma, magnets and magnetic field.
2. ABJEG is easier, cheaper by ten (perhaps a hundred) times and more efficient.
3. ABJEG may be used for getting energy from wind, river and moving water (ocean stream).
4. ABJEG can be small and it may be used for getting energy in small vehicles.
5. ABJEG can work as propulsion or pump.

Advantages ABJEG over the conventional electric (magnetic) generator:
1. ABJEG is easier, by some times than a conventional generator.
2. ABJEG produces high voltage electricity. Big electric stations do not need a heavy expensive and vulnerable transformer.
3. ABJEG produces a direct current (DC). That is suitable for transfer over long distance.
4. The small DC generators can easy connect to the common net. Not necessary the harmonize the frequency and phase of current.

Theory of Jet Electric generator. Computation and Estimation.
1. Ion and electron speed.
  Ion mobility. The ion speed onto the gas (air) jet may be computed by equation:
$$j_s = qn_-b_-E + qD_-(dn_-/dx), \qquad (1)$$
where j_s is density of electric current about jet, A/m^2; $q = 1.6 \times 10^{-19}$ C is charge of single electron, C; n_- is density of injected negative charges in 1 m^3; b_- is charge mobility of negative charges, m^2/sV; E is electric intensity, V/m; D_- is diffusion coefficient of charges; dn_-/dx is gradient of charges. For our estimation we put $dn_-/dx = 0$. In this case
$$j_s = qn_-b_-E, \quad Q = qn_-, \quad v = bE, \quad j_s = Qv, \qquad (2)$$
where Q is density of the negative charge in 1 m^3; v is speed of the negative charges about jet, m/s.
  The air negative charge mobility for normal pressure and temperature $T = 20°C$ is:
  In dry air $b_- = 1.9 \times 10^{-4}$ m^2/sV, in humid air $b_- = 2.1 \times 10^{-4}$ m^2/sV. (3)
In Table 1 is given the ions mobility of different gases for pressure 700 mm Hg and for $T = 18°C$.

Table 1. Ions mobility of different gases for pressure 700 mm Hg and for $T = 18$ °C.

Gas	Ion mobility 10^{-4} m^2/sV, b_+, b_-	Gas	Ion mobility 10^{-4} m^2/sV, b_+, b_-	Gas	Ion mobility 10^{-4} m^2/sV, b_+, b_-
Hydrogen	5.91 8.26	Nitrogen	1.27 1.82	Chloride	0.65 0.51
Oxygen	1.29 1.81	CO$_2$	1.10 1.14		

Source [8] p.357.

In diapason of pressure from 13 to 6×10^6 Pa the mobility follows the Law bp = const, where p is air pressure. When air density decreases, the charge mobility increases. The mobility strength depends upon the purity of gas. The ion gas mobility may be recalculated in other gas pressure p and temperature T by equation:

$$b = b_0 \frac{T p_0}{T_0 p}, \quad (4)$$

where lower index "$_0$" mean the initial (known) point. At the Earth surface $H = 0$ km, $T_0 = 288$ K, $p = 1$ atm; at altitude $H = 10$ km, $T_0 = 223$ K, $p = 0.261$ atm;

For normal air density the electric intensity must be less than 3 MV ($E < 3$ MV/m) and depends from pressure.

Electron mobility. The ratio $E/p \approx$ constant. Conductivity σ of gas depends upon density of charges particles n and their mobility b, for example:

$$\sigma = neb, \quad \lambda = 1/n\sigma, \quad (5)$$

where b is mobility of the electron, λ is a free path of electron.

Electron mobility depends from ratio E/n. This ratio is given in Table 2.

Table 2. Electron mobility b_e in gas vs E/n

Gas	$E/n \times 10^{-17}$ 0.03 V·cm²	$E/n \times 10^{-17}$ 1 V·cm²	$E/n \times 10^{-17}$ 100 V·cm²	Gas	$E/n \times 10^{-17}$ 0.03 V·cm²	$E/n \times 10^{-17}$ 1 V·cm²	$E/n \times 10^{-17}$ 100 V·cm²
N_2	13600	670	370	He	8700	930	1030
O_2	32000	1150	590	Ne	16000	1400	960
CO_2	670	780	480	Ar	14800	410	270
H_2	5700	700	470	Xe	1980	-	240

Source: Physics Encyclopedia http://www.femto.com.ua/articles/part_2/2926.html

The electrons may connect to the neutral molecules and produce the negative ions (for example, affinity of electron to O_2 equals 0.3 - 0.87 eV, to H_2O equals 0.9 eV [7] p.424). That way the computation of the mobility of a gas containing electrons and ions is a complex problem. Usually the computations are made for all electrons converted to ions.

The maximal electric intensity in air at the Earth surface is $E_m = 3$ MV/m. If atmospheric pressure changes the E_m also changes by law $E_m/p =$ constant.

Example 4. If $E = 10^5$ V/m, than $v = 20$ m/s in Earth surface conditions.

2. Electron injectors.

There are some methods for getting the electron emissions: hot cathode emission, cold field electron emission (edge cold emission, edge cathode). The photo emission, radiation emission, radioisotope emission and so on usually produce the positive and negative ions together. We consider only the hot emission and the cold field electron emission (edge cathodes), which produces only electrons.

Hot electron emission.

Current i of diode from potential (voltage) U is

$$i = CU^{3/2} \quad (6)$$

where C is constant which depends from form and size cathode. For plate diode

$$C = \frac{4}{9} \varepsilon_0 \frac{S}{d^2} \sqrt{\frac{2e}{m_e}} \approx 2.33 \cdot 10^{-6} \frac{S}{d^2}, \quad (7)$$

where $\varepsilon_o = 8.85 \cdot 10^{-12}$ F/m; S is area of cathode (equals area of anode), cm²; d is distance between cathode and anode, cm; e/m_e is the ratio of the electron charge to electron mass, C/kg;

Result of computation equation (7) is in fig. 5.

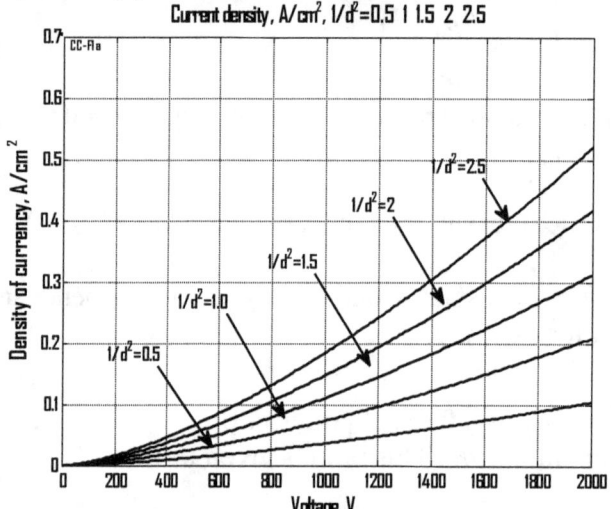

Fig.5. Electric current via voltage the plain cathodes for different ratio of the distance.

The maximal **hot cathode** emission computed by equation:

$$j_s = BT^2 exp(-A/kT), \qquad (8)$$

where B is coefficient, A/cm²K²; T is cathode temperature, K; $k = 1.38 \times 10^{-23}$ [J/K] is Bolzmann constant; $A = e\varphi$ is thermoelectron exit work, J ; φ is the exit work (output energy of electron) in eV, $e = 1.6 \cdot 10^{-19}$. Both values A, B depend from material of cathode and its cover. The "A" changes from 1.3 to 5 eV, the "B" changes from 0.5 to 120 A/cm²K². Boron thermo-cathode produces electric current up 200 A/cm². For temperature 1400 -1500K the cathode can produce current up 1000 A/cm². The life of cathode can reach some years.

Exit energy from metal are (eV):
$$\text{W 4.5, Mo 4.3, Fe 4.3, Na 2.2 eV}, \qquad (9)$$
From cathode covered by optimal layer(s) the exit work is in Table 3.

Table 3. Exit work (eV) from cathode is covered by the optimal layer(s):

Cr – Cs	Ti – Cs	Ni – Cs	Mo – Cs	W – Ba	Pt – Cs	W - O – K	Steel- Cs	Mo₂C- Cs	WSi₂- Cs
1.71	1.32	1.37	1.54	1.75	1.38	1.76	1.52	1.45	1.47

Source [8]: Kikoin, Table of physic values, 1976, p. 445 (in Russian).

Results of computation the maximal electric current (in vacuum) via cathode temperature for the different exit work of electrons f are presented in fig.6.

Method of producing electrons and positive ions is well developed in the ionic thrusters for space apparatus.

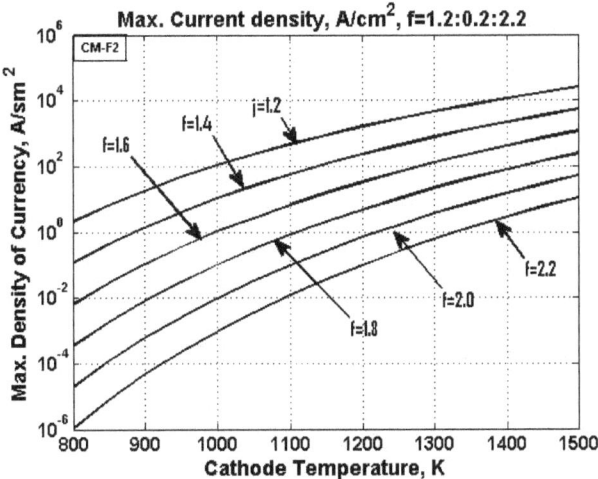

Fig.6. The maximal electric current via cathode temperature for the different exit work of electrons f.

The field electron emission. (The edge cold emission).

The cold field electron emission uses the edge cathodes. It is known that the electric intensity E_e in the edge (needle) is

$$E_e = U/a. \qquad (10)$$

Here a is radios of the edge. If voltage between the edge and nears net (anode) is $U = 1000$ V, the radius of edge $a = 10^{-5}$ m, electric intensity at edge is the $E_e = 10^8$ V/m. That is enough for the electron emission. The density of electric current may reach up 10^4 A/cm^2. For getting the required current we make the need number of edges.

The density of electric current approximately is computed by equation:

$$j \approx 1.4 \cdot 10^{-6} \frac{E^2}{\varphi} 10^{(4.39\varphi^{-1/2} - 2.82 \cdot 10^7 \varphi^{3/2}/E)}, \qquad (11)$$

where j is density of electric current, A/cm^2; E is electric intensity near edge, V/cm; φ is exit work (output energy of electron, field electron emission), eV.

The density of current is computed by equation (11) in Table 4 below.

$\varphi = 2{,}0$ eV		$\varphi = 4{,}5$ eV		$\varphi = 6{,}3$ eV	
$E \times 10^{-7}$	lg j	$E \times 10^{-7}$	lg j	$E \times 10^{-7}$	lg j
1,0	2,98	2,0	-3,33	2,0	-12,9
1,2	4,45	3,0	1,57	4,0	-0,88
1,4	5,49	4,0	4,06	6,0	3,25
1,6	6,27	5,0	5,59	8,0	5,34
1,8	6,89	6,0	6,62	10,0	6,66
2,0	7,40	7,0	7,36	12,0	7,52
2,2	7,82	8,0	7,94	14,0	8,16
2,4	8,16	9,0	8,39	16,0	8,65
2,6	8,45	10,0	8,76	18,0	9,04
		12,0	9,32	20,0	9,36

Source: http://www.femto.com.ua/articles/part_1/0034.html

Example: Assume we have needle with edge $S_1 = 10^{-4}$ cm^2, $\varphi = 2$ eV and net $S_2 = 10 \times 10 = 10^2$ cm^2

located at distance $L = 10$ cm. The local voltage between the needle and net is $U = 10^2$ volts. Than electric intensity at edge of needle, current density and the electric current is:

$$E = \frac{S_2 U}{S_1 L} = \frac{10^2 10^2}{10^{-4} 10^1} = 10^7 \text{ V/cm}, \quad j = 10^3 \text{ A/cm}^2, \quad i = jS_1 = 10^3 10^{-4} = 0.1 \text{ A}, \quad (12)$$

Here j is taken from Table 4 or computed by equation (11). If we need in the electric current 10 A, we must locate 100 needles in the entrance area 1×1 m of generator.

Computation of equation (11) is presented in fig. 7.

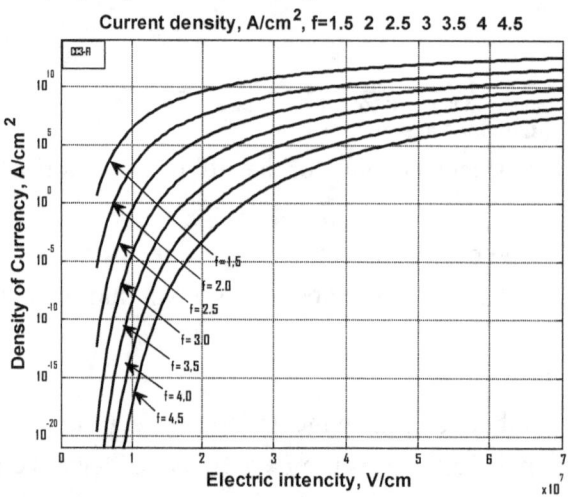

Fig.7. Density of electric current the noodle injector via the electric intensity for different the field electron emissions f.

3. Internal and outer pressure on the generator surface.

The electric charges located in the ABJEG generator produce electric intensity and internal and outer pressure. The electric intensity can create electrical breakdown; the pressure can destroy the generator.
a) For the cylindrical generator the electric intensity and pressure may be estimated by equations:

$$E = k\frac{2\tau}{\varepsilon r}, \quad \tau = \frac{i}{V_a}, \quad p = E\sigma, \quad (13)$$

where E is electric intensify, V/m; $k = 9 \cdot 10^9$ is electric constant, Nm^2/C^2; τ is the linear charge, C/m; ε is dielectric constant for given material ($\varepsilon = 1 - 1000$), r is radius of tube (generator), m; i is electric current A; V_a is average speed of flow inside of generator, m/s; p is pressure, N/m^2; σ is the density of charge, C/m^2 at an tube surface.

Example. Assume the generator has $r = 0.1$ m, $V_a = 50$ m/s, $i = 0.1$ A. Let us take as isolator GE Lexan having the dielectric strength $E_m = 640$ MV/m and $\varepsilon = 3$. Than from (13) we have $E = 120$ MV/m $< E_m = 640$ MV/m.

If $E > E_m$ we can locate the part of the compensate charge inside generator.
b) For plate of generator having rectangular entrance h×w = 1× 3 m and compensation charges on two sides, the electric intensity and pressure may be estimated by equations:

$$E = 4\pi k \frac{\sigma}{\varepsilon}, \quad \tau = \frac{i}{2V_a w}, \quad p = 2\varepsilon\varepsilon_0 E^2, \quad (14)$$

where w is width of entrance, m; ε is dielectric coefficient of the isolator.

4. Loss of energy and matter to ionization.

Let us estimate the energy and matter requested for ionization and discharge in the offered ABJEG generator. Assume we have ABJEG generator having the power $P = 10,000$ kW and a work voltage $V = 1$ MV. In this case the electric current is $i = P/V = 10$ A $= 10$ C/s.

Assume we use the nitrogen N_2 for ionization (a very bad gas for it). It has exit work about 5 eV and relative molecular weight 14. One molecule (ion) of N_2 weights $m_N = 14 \cdot 1.67 \cdot 10^{-27} = 2.34 \cdot 10^{-26}$ kg. The 1 ampere has $n_A = 1/e = 1/1.6 \cdot 10^{-19} = 6.25 \cdot 10^{18}$ ions/s. Consumption of the ion mass is:
$M = m_{Ni} \, n_A = 2.34 \cdot 10^{-26} \cdot 10 \cdot 6.25 \cdot 10^{18} = 1.46 \cdot 10^{-6}$ kg/s $= 1.46 \cdot 10^{-6} \cdot 3.6 \cdot 10^{3} = 5.26 \cdot 10^{-3}$ kg/hour ≈ 5 gram/hour.

If electron exit work equals $\varphi = 4.5$ eV the power spent extraction of one electron is: $E_1 = \varphi e = 4.5 \cdot 1.6 \cdot 10^{-19} = 7.2 \cdot 10^{-19}$ J.

The total power for the electron extraction is $E = i \cdot n_A \cdot E_1 = 10 \cdot 6.25 \cdot 10^{18} \cdot 7.2 \cdot 10^{-19} = 45$ W.

The received values mass M and power E are very small in comparison with conventional consumption of fuel (tons in hour) and generator power (thousands of kW).

Important note (Compensation of flow charge). Any contact collector cannot collect ALL charges. Part of them will fly away. That means the generator (apparatus) will be charged positive (if fly away electrons or negative ions) or negative (if fly away the positive ions). It is easy to delete the negative charges by edge. The large positive charge we may delete by a small ion accelerator. The art of ion engines for vacuum is well developed. They may be used as injectors and dischargers in the first design.

The charges may be deleted also by grounding.

Below is spark gap in air.

Table 5. Electric spark in air (in mm. For normal atmospheric pressure).

Voltage, kV	Two edges,	Two spheres, $d = 5$ sm	Two plates
20	15.5	5.8	6.1
40	45.5	13	13.7
100	200	15	36.7
		Two spheres, $d = 2$ sm	
200	410	262	75.3
300	600	530	114

Source [6], p124.

Application of Jet Electric Generator

1. Electric Station.

Estimations of main parameters of ABJEG for an electric station.

Assume we have ABJEG as the cylindrical tube with constant cross-section area $f = 0.01$ m^2 (fig.8). Let us take the pressure in balloon 1 $p = 0.5$ atm $= 50,000$ N/m^2 and the gas (air) speed in exit of ABJEG $V = 50$ m/s.

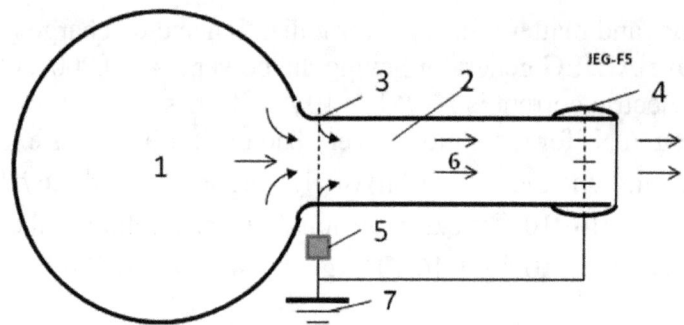

Fig.8. Principal schema of the Jet Electric Generator. *Nominations:* 1 – balloon with pressure gas, 2 – jet electric generator (ABJEG), 3 – injector of electrons, 4 – collector of electrons, 5 – outer load, 6 – gas flow, 7 – grounding.

The useful (working) pressure equals the pressure p into balloon 1 minus the kinetic loss of a gas in the exit

$$\Delta p_1 = p - \frac{\rho V^2}{2} = 5 \cdot 10^4 - \frac{1.225 \cdot 50^2}{2} = 4.847 \cdot 10^4, \quad \frac{N}{m^2}. \quad (15)$$

That is anti-pressure of electron (ions) moving against the flow.
Power of ABJEG is

$$P = f \cdot \Delta p_1 V = 0.01 \cdot 4.847 \cdot 10^4 \cdot 50 = 24.2 \text{ kW}, \quad (16)$$

Coefficient of efficiency of ABJEG

$$\eta = \frac{\Delta p_1}{p} = \frac{4.847}{5} = 0.97. \quad (17)$$

Let us estimate the voltage and current for length $L = 0.3$ m of active part of tube.
The maximum of electric intensity E_m must be less than

$$E_m < \frac{V}{b} = \frac{50}{2 \cdot 10^{-4}} = 2.5 \cdot 10^5 \frac{V}{m}. \quad (18)$$

Let us take the electric intensity $E = 2 \cdot 10^5$ V/m. Than the work voltage will be

$$U = EL = 2 \cdot 10^5 \cdot 0.3 = 60 \text{ kV}. \quad (19)$$

The current will be

$$i = P/U = 24.2/60 = 0.4 \text{ A}. \quad (20)$$

The ABJEG is suitable as simple (tube) additional electric generator for internal combustion engines working by Otto's cycle or any machine having pressure or high speed exhaust gases. That increases their efficiency. Vast industrial possibilities are opened by recovery of otherwise waste energies at low opportunity cost.

The under critical speed w and consumption m of ideal gases from the converging nozzle may be estimated by equations:

$$w = \sqrt{2\frac{k}{k-1}p_1 v_1 \left[1 - \left(\frac{p_2}{p_1}\right)^{\frac{k-1}{k}}\right]}, \quad m = f\sqrt{2\frac{k}{k-1}\frac{p_1}{v_1}\left[\left(\frac{p_2}{p_1}\right)^{2/k} - \left(\frac{p_2}{p_1}\right)^{\frac{k+1}{k}}\right]}, \quad (21)$$

where k is adiabatic coefficient in gas dynamic; p is gas pressure in beginning "1" and end "2" of nozzle; v is specific gas volume, f is cross-section area of tube, m².

Critical ratio β_k and critical speed w_k is

$$\beta_k = \frac{p_k}{p_1} = \left(\frac{2}{k+1}\right)^{\frac{k}{k-1}}, \quad \beta_k < \frac{p_2}{p_1} < 1, \quad w_k = \sqrt{2\frac{k}{k+1}p_1 v_1}. \quad (22)$$

For one atom gas $k = 1.66$ and $\beta_k = 0.42$,
For two atom gas $k = 1.4$ and $\beta_k = 0.528$,
For one atom gas $k = 1.3$ and $\beta_k = 0.546$.

Equation of gas state and continuity equation is

$$pV = mRT, \quad \frac{fw}{v} = const. \quad (23)$$

Here V is gas volume, m³; m is gas mass, kg; T is gas temperature, K; R is gas constant, for air $R = 287$ J/kg K.
These equations allow to compute the data of gas flow.

The steam is very suitable gas for converting its extension directly to electricity without steam piston machines, steam turbines and magnetic generator directly to electricity by ABJEG.
The steam speed after a converting nozzle may be computed by equation

$$w = 44.72\sqrt{i_1 - i_2}, \quad (24)$$

where I is the steam enthalpy in the beginning and end of the adiabatic process. The enthalpy is found in diagram "is" by data of the beginning and end of the adiabatic process.

2. Wind Energy.

The simplest wind electric generator (ABJEG) is shown in fig.9. In end of mast 1 is installed the electron injector 2. The wind 6 pick up the electrons and moves them to the Earth surface. Under Earth surface the electros throw the grounding 4 and outer electric load 3 return to injector 2.

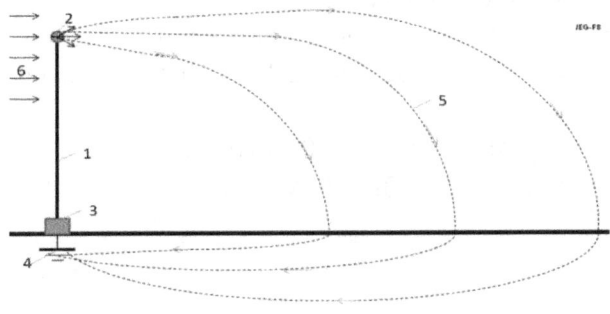

Fig.9. Simplest wind generator. *Notations:* 1 – mast, 2 - electron injector, 3 – electric load, 4 – grounding, 5 – trajectories of electrons, 6 = wind.

The electric resistance of the grounding may be estimated by equation

$$R = \frac{\rho}{2\pi a}, \quad (25)$$

ρ is specific average resistance of the ground, Ωm; a is average radius the plate of grounding. The good grounding must be in place of underground water or in sea water. For sea water $\rho \approx 0.2 \, \Omega m$. For underground water ρ is below. Connection having one line underground widely uses in communication.

Suggested method may be used for getting the wind energy at high altitude. The injector must be supported at high altitude by balloon, dirigible or wing apparatus [1].

a). **Power of a wind** energy N [Watt, Joule/sec]

$$N = 0.5 \eta \rho A V^3 \quad [W]. \quad (26)$$

The coefficient of efficiency, η, equals about 0.2 -0.25 for EABG; 0.15 - 0.35 for low speed propeller rotors (ratio of blade tip speed to wind speed equals $\lambda \approx 1$); $\eta = 0.45 - 0.5$ for high speed propeller rotors ($\lambda = 5 - 7$). The Darrieus rotor has $\eta = 0.35 - 0.4$. The gyroplane rotor has 0.1 - 0.15. The air balloon and the drag (parachute) rotor has $\eta = 0.15 - 0.2$. The Makani rotor has 0.15 - 0.25. The theoretical maximum equals $\eta \approx 0.6$. Theoretical maximum of the electron generator is 0.25. A - front (forward) area of the electron injector, rotor, air balloon or parachute [m²]. ρ - density of air: $\rho_0 = 1.225$ kg/m³ for air at sea level altitude $H = 0$; $\rho = 0.736$ at altitude $H = 5$ km; $\rho = 0.413$ at $H = 10$ km. V is average annually wind speed, m/s.

Table 5. Relative density ρ_r and temperature of the standard atmosphere via altitude

H, km	0	0.4	1	2	3	6	8	10	12
$\rho_r = \rho/\rho_0$	1	0.954	0.887	0.784	0.692	0.466	0.352	0.261	0.191
T, K	288	287	282	276	269	250	237	223	217

Source [6].

The salient point here is that the strength of wind power depends upon the wind speed (by third order!). If the wind speed increases by two times, the power increases by 8 times. If the wind speed increases 3 times, the wind power increases 27 times!

The wind speed increases in altitude and can reach in constant air stream at altitude $H = 5 - 7$ km up $V = 30 - 40$ m/s. At altitude the wind is more stable/constant, which is one of the major advantages that an airborne wind systems has over ground wind systems.

For comparison of different wind systems of the engineers must make computations for average annual wind speed $V_0 = 6$ m/s (or 10 m/s) and altitude $H_0 = 10$ m. For standard wind speed and altitude the maximal wind power equals 66 W/m².

The energy, E, produced in one year is (1 year $\approx 30.2 \times 10^6$ work sec) [J]

$$E = 3600 \times 24 \times 350 N \approx 30 \times 10^6 N, \quad [J]. \quad (27)$$

3. Water Energy.

Typical hydroelectric station is shown in fig. 10. The water from a top level 1 flows by tube 2 to ABJEG 3 and after working runoff to lower level 4.

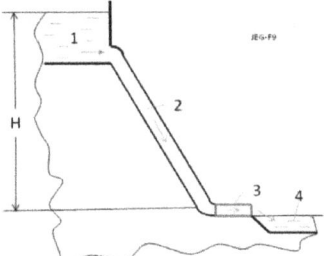

Fig. 10. Typical Hydroelectric station with ABJEG. *Notations:* 1 - upper source; 2 - water canal; 3 – ABJEG; 4 – lower runoff.

Note: It is possible also the water electric generator shown in fig. 8. One may be used in rivers and ocean streams.

1. **Power of a water flow** is N [Watt, Joule/sec]:

$$N = \eta \rho B g H \quad [W]. \tag{28}$$

The coefficient of efficiency, η, equals about 0.8 - 0.95; ρ - density of liquid: $\rho \approx 1.000$ kg/m^3 for water; B is the flow in cubic meters per second; $g = 9.81$ m/s^2 is Earth gravity; H is the height difference between inlet and outlet of installation, m (fig.10).

Without ABJEG the H and V connected by equation

$$H = V^2/2g \ . \tag{29}$$

2. Resistance of water. Salt water conducts an electric current. The specific electric resistance of water is significantly depends from salinity of water. When we have the plates (nets) with both sides (cathode and anode), the specific electric resistance are:
  1. Distilled water $R \approx 10^6$ Ωm.
  2. Fresh water $R = 40 - 200$ Ωm (depends from water salinity). (30)
  3. Sea water $R \approx 0.2$ Ωm.

In our case in one side we have the electron injector (cathode) which has conventionally a small area. In this case the specific electric resistance is:

$$R_o = R/ 4\pi a \ , \tag{31}$$

where a is radius of needle (or cathode), m; this radius conventionally is very small (mm). That means the R_o has an electric resistance of hundreds Ohms. We can neglect their influence in the installation efficiency .

3. Electron speed in water.
The charge mobility in water is:
$$Cl^- \text{ is } 0.667 \times 10^{-7} \text{ m}^2/\text{sV}, \quad Na^+ \text{ is } 0.450 \times 10^{-7} \text{ m}^2/\text{sV}. \tag{32}$$

As you see the mobility of ions in water is very small. The applied voltage in water is also small. That means the ion speed v is small in the comparison with water speed. In many case we can put $v = bE \approx 0$

If $v > 0$, the electrons accelerate the water ($E > 0$ and installation spends energy, works as engine). If $v < 0$, the electrons brake the water ($E < 0$ and the correct installation can produce energy, works as electric generator). If $v = 0$ (electron speed about installation equals water speed V), the electric resistance is zero.

4. The efficiency of installation from back electric current may be estimated by equation:
$$\eta \approx 1/(1+ R_u/R_o) \,, \tag{33}$$
where R_u is an useful electric resistance. Ratio R_u/R_o conventionally is small and η is closed to 1.

5. Specific power of Installation N_1 [W/m²]. The specific power of the offered installation may be estimated by a series of equations:
$$N_1 \approx \eta\, A_1/t = \eta Q_1 EL/t = \eta Q_1 EV = j_s U = \eta \rho B_1 g h = 0.5 \eta m_1 V^2, \tag{34}$$
where A_1 is energy of flow through 1 m², J/m²; t is time, sec; B_1 is flow in m³ through cross section area of flow 1 m²; E is electric intensity, V/m; L is distance between injector and net (cathode and anode); V is flow speed, m/s; j_s is density of electric **current**, A/m²; U is electric voltage, V; m_1 is flow mass per second through area 1 m²; Q_1 is density of the negative charge in 1 m³; $g = 9.81$ m/s² is Earth gravity; h is the height difference between inlet and outlet of installation (between electron injector and net, between cathode and anode), m.

Summary

Relatively no progress has been made in electric generators in the last years.

The author proposes a fundamentally new efficient electric generator for gas and liquid. No gas (water) turbine, no dynamo-machine. Practically there is only a tube.

It is not comparable to conventional MHD generator or heat machine. The MHD generator requests very high temperature, which cannot be endured by available materials. MHD generator is very complex and expensive.

Author offers and develops theory of a new simple cheap and efficient electric (electron) generator. This generator can convert pressure or kinetic energy of the any non-conductive flow (gas, liquid) into direct current (DC). The generator can convert the mechanical energy of any engine in high voltage DC. One can covert in electricity the wind and water energy without turbine. One can convert the rest energy of an internal combustion engine or turbojet engine in electricity and increase its efficiency. ABJEG may be propulsors, which have been applied to pump a gas or dielectric liquid and as engine in several experimental ships.

As any new idea, the suggested concept is in need of research and development. The theoretical problems do not require fundamental breakthroughs. It is necessary to design small, cheap installations to study and get an experience in the design electron wind (water) generator.

This paper has suggested some design solutions from patent application. The author has many detailed analysis in addition to these presented projects. Organizations or investors are interested in these projects can address the author (http://Bolonkin.narod.ru , aBolonkin@juno.com , abolonkin@gmail.com).

The closed ideas are in [1]-[5]. Researches and information related to this topic are presented in [6]-[9].

ACKNOWLEDGEMENT

The author wishes to acknowledge Joseph Friedlander for correcting the English and offering useful advice and suggestions.

References

[1] Bolonkin A.A., **Electronic Wind Generator**.
Electrical and Power Engineering Frontier Sep. 2013, Vol. 2 Iss. 3, PP. 64-71.

http://www.academicpub.org/epef/Issue.aspx?Volume=2&Number=3&Abstr=false
http://viXra.org/abs/1306.0046 , www.IntellectualArchive.com,
https://archive.org/details/ArticleElectronWindGenerator6613AsterShmuelWithPicture
http://www.scribd.com/doc/146177073/Electronic-Wind-Generator

[2] Bolonkin A.A., **Electron Hydro Electric Generator.** International Journal of Advanced Engineering Applications. ISSN: 2321-7723 (Online), Special Issue I, 2013.
http://fragrancejournals.com/?page_id=18, http://viXra.org/abs/1306.0196,
http://www.scribd.com/doc/149489902/Electron-Hydro-Electric-Generator , #1089
http://archive.org/details/ElectronHydroElectricGenerator_532, http://intellectualarchive.com,

[3] Bolonkin A.A., **Electron Super Speed Hydro Propulsion.** International Journal of Advanced Engineering Applications, Special Issue 1, pp.15-19 (2013), http://viXra.org/abs/1306.0195
http://www.scribd.com/doc/149490731/Electron-Super-Speed-Hydro-Propulsion
http://archive.org/details/ElectronSuperSpeedHydroPropulsion
http://intellectualarchive.com . #1090
http://fragrancejournals.com/wp-content/uploads/2013/03/Special-Issue-1-4.pdf

[4] Bolonkin A.A., **Electron Air Hypersonic Propulsion**. International Journal of Advanced Engineering Applications, Vol.1, Iss. 6, pp.42-47 (2012). http://viXra.org/abs/1306.0003,
http://www.scribd.com/doc/145165015/Electron-Air-Hypersonic-Propulsion ,
http://www.scribd.com/doc/146179116/Electronic-Air-Hypersonic-Propulsion ,
http://fragrancejournals.com/wp-content/uploads/2013/03/IJAEA-1-6-6.pdf .

[5] Bolonkin A.A., **Electric Hypersonic Space Aircraft.** Global Science Journal, 2 July, 2014,
http://viXra.org/abs/1407.0011, http://www.scribd.com/doc/232209230.
http://intellectualarchive.com Ref. #1288

[6] N.I. Koshkin and M.G. Shirkebich, Directory of Elementary Physics, Nauka, Moscow, 1982 (in Russian).

[7] I.K. Kikoin. Table of Physics values. Atomisdat, Moscow, 1976 (in Russian).

[8] S.G. Kalashnikov, Electricity, Moscow, Nauka, 1985.(in Russian).

[9] Wikipedia. Electric Generator, http://wikipedia.org .

May 27, 2014

Chapter 5

Wireless Transfer of Electricity from Continent to Continent
Alexander Bolonkin

Abstract

Author offers collections from his previous research of the revolutionary new ideas: wireless transferring electric energy in long distance – from one continent to other continent through Earth ionosphere and storage the electric energy into ionosphere. Early he also offered the electronic tubes as the method of transportation of electricity into outer space and the electrostatic space 100 km towers for connection to Earth ionosphere.

Early it is offered connection to Earth ionosphere by 100 km solid or inflatable towers. There are difficult for current technology. In given work the research this connection by thin plastic tubes supported in atmosphere by electron gas and electrostatic force. Building this system is cheap and easy for current technology.

The computed project allows estimating the possibility of the suggested method.

Key words: transferring of electricity in space; transfer of electricity to spaceship, Moon, Mars; plasma MagSail; electricity storage; ionosphere transfer of electricity.

Introduction

The production, storage, and transference of large amounts of electric energy is an enormous problem for humanity. These spheres of industry are search for, and badly need revolutionary ideas. If in production of energy, space launch and flight we have new ideas (see [1]-[15]), the new revolutionary ideas in transferring and storage energy are only in the works [1-6].

Important Earth mega-problem is efficient transfer of electric energy long distances (intra-national, international, intercontinental). The consumption of electric energy strongly depends on time (day or night), weather (hot or cold), from season (summer or winter). But electric station can operate most efficiently in a permanent base-load generation regime. We need to transfer the energy a far distance to any region that requires a supply in any given moment or in the special hydro-accumulator stations. Nowadays, a lot of loss occurs from such energy transformation. One solution for this macro-problem is to transfer energy from Europe to the USA during nighttime in Europe and from the USA to Europe when it is night in the USA. Another solution is efficient energy storage, which allows people the option to save electric energy.

The storage of a big electric energy can help to solve the problem of cheap space launch. The problem of an acceleration of a spaceship can be solved by use of a new linear electrostatic engine suggested in [10] or Magnetic Space Launcher offered in [11]. However, the cheap cable space launch offered by author [12] requires use of gigantic energy in short time period. (It is inevitable for any launch method because we must accelerate big masses to the very high speed - 8 ÷11 km/s). But it is impossible to turn off whole state and connect all electric station to one customer. The offered electric energy storage can help solving this mega-problem for humanity.

The idea of wireless transfer energy through ionosphere was offered and researched by author in [1 - 6]. For connection to Earth ionosphere offered the 100 km solid, inflatable, electrostatic or kinetic towers [7 - 9]. But it is expensive and difficult for current technology.

Wireless transferring of electric energy in Earth.

It is interesting the idea of energy transfer from one Earth continent to another continent without wires. As it is known the resistance of infinity (very large) conducting medium does not depend from distance. That is widely using in communication. The sender and receiver are connected by only one wire, the other wire is Earth. The author offers to use the Earth's ionosphere as the second plasma cable. It is known the Earth has the first ionosphere layer E at altitude about 100 km (Fig. 1). The concentration of electrons in this layer reaches 5×10^4 1/cm³ in daytime and 3.1×10^3 1/cm³ at night (Fig. 1). This layer can be used as a conducting medium for transfer electric energy and communication in any point of the Earth. We need minimum two space 100 km. towers (Fig. 2). The cheap optimal inflatable, kinetic, and solid space towers are offered and researched by author in [6-9]. Additional innovations are a large inflatable conducting balloon at the end of the tower and big conducting plates in a sea (ocean) that would dramatically decrease the contact resistance of the electric system and conducting medium.

Theory and computation of these ideas are presented in Macroprojects section.

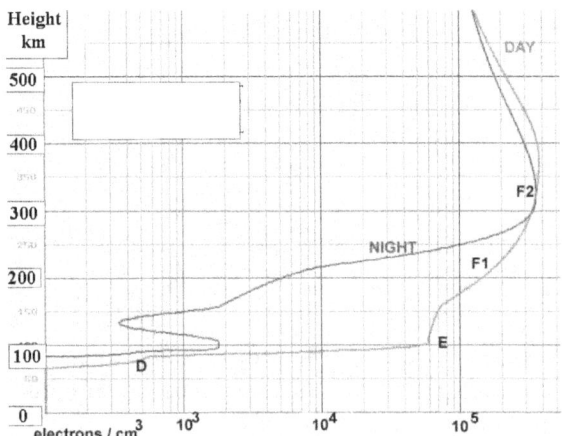

Fig.1. Consentration/cm³ of electrons (= ions) in Earth's atmosphere in the day and night time in the D, E, F1, and F2 layers of ionosphere.

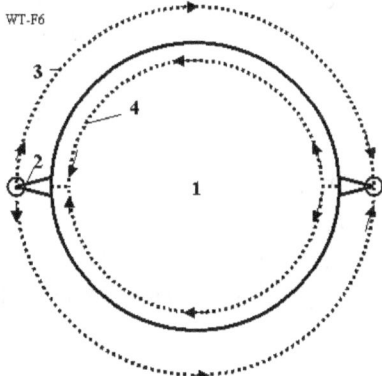

Fig.2. Using the ionosphere as conducting medium for transferring a huge electric energy between continents and as a large storage of the electric energy. Notations: 1 - Earth, 2 - space tower (or

electron tube) about 100 km of height, 3 - conducting *E* layer of Earth's ionosphere, 4 - back connection through Earth.

However the solid 100 km space towers are very expensive. Main innovation in this work is connection to ionosphere by cheap film tube filled by electron gas.

Electronic tubes

The author's first innovations in electrostatic applications were developed in 1982-1983 [1]-[3].

Later the series articles of this topic were published in [4]-[15]. In particular, in the work [4-5] was developed theory of electronic gas and its application to building (without space flight!) inflatable electrostatic space tower up to the stationary orbit of Earth's satellite (GEO).

In given work this theory applied to special inflatable electronic tubes made from thin insulator film. It is shown the charged tube filled by electron gas is electrically neutral, that can has a high internal pressure of the electron gas.

The main property of AB electronic tube is a very low electric resistance because electrons have small friction on tube wall. (In conventional solid (metal) conductors, the electrons strike against the immobile ions located in the full volume of the conductor.). The abnormally low electric resistance was found along the lateral axis only in nanotubes (they have a tube structure!). In theory, metallic nanotubes can have an electric current density (along the axis) more than 1,000 times greater than metals such as silver and copper. Nanotubes have excellent heat conductivity along axis up 6000 W/m·K. Copper, by contrast, has only 385 W/m·K. The electronic tubes explain why there is this effect. Nanotubes have the tube structure and electrons can free move along axis (they have only a friction on a tube wall).

More over, the moving electrons produce the magnetic field. The author shows - this magnetic field presses against the electron gas. When this magnetic pressure equals the electrostatic pressure, the electron gas may not remain in contact with the tube walls and their friction losses. The electron tube effectively becomes a superconductor for any surrounding temperature, even higher than room temperature! Author derives conditions for it and shows how we can significantly decrease the electric resistance.

Description, Innovations, and Applications of Electronic tubes.

An electronic AB-Tube is a tube filled by electron gas (fig.3). Electron gas is the lightest gas known in nature, far lighter than hydrogen. Therefore, tubes filled with this gas have the maximum possible lift force in atmosphere (equal essentially to the lift force of vacuum). The applications of electron gas are based on one little-known fact – the electrons located within a cylindrical tube having a positively charged cover (envelope) are in neutral-charge conditions – the total attractive force of the positive envelope plus negative contents equals zero. That means the electrons do not adhere to positive charged tube cover. They will freely fly into an AB-Tube. It is known, if the Earth (or other planet) would have, despite the massive pressures there, an empty space in Earth's very core, any matter in this (hypothetical!) cavity would be in a state of weightlessness (free fall). All around, attractions balance, leaving no vector 'down'.

Analogously, that means the AB-Tube is a conductor of electricity. Under electric tension (voltage) the electrons will collectively move without internal friction, with no vector 'down' to the walls, where friction might lie. In contrast to movement of electrons into metal (where moving electrons impact against a motionless ion grate). In the AB-Tube we have only electron friction about the tube wall. This friction is significantly less than the friction electrons would experience against ionic structures—and therefore so is the electrical resistance.

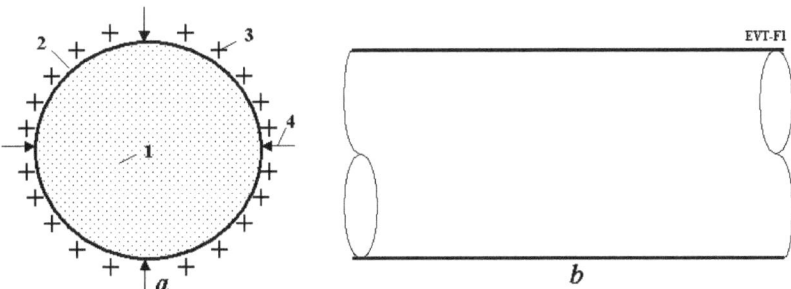

Fig.3. Electronic vacuum AB-Tube. *a*) Cross-section of tube. *b*) Side view. *Notation*: 1 – Internal part of tube filled by free electrons; 2 – insulator envelope of tube; 3 – positive charges on the outer surface of envelope (over this may be an additional film-insulator); 4 – atmospheric pressure.

When the density of electron gas equals $n = 1.65 \times 10^{16}/r$ $1/m^3$ (where r is radius of tube, m), the electron gas has pressure equals atmospheric pressure 1 atm (see research below). In this case the tube cover may be a very thin—though well-sealed-- insulator film. The outer surface of this film is charged positively by static charges equal the electron charges and AB-Tube is thus an electrically neutral body.

Moreover, when electrons move into the AB-Tube, the electric current produces a magnetic field (fig.4). This magnetic field compresses the electron cord and decreases the contact (and friction, electric resistance) electrons to tube walls. In the theoretical section is received a simple relation between the electric current and linear tube charge when the magnetic pressure equals to electron gas pressure $i = c\tau$ (where i is electric current, A; $c = 3 \times 10^8$ m/s – is the light speed; τ is tube linear electric charge, C/m). In this case the electron friction equals zero and AB-Tube becomes **superconductive at any outer temperature**. Unfortunately, this condition requests the electron speed equals the light speed. It is, however, no problem to set the electron speed very close to light speed. That means we can make the electric conductivity of AB-Tubes very close to superconductivity almost regardless of the outer temperature.

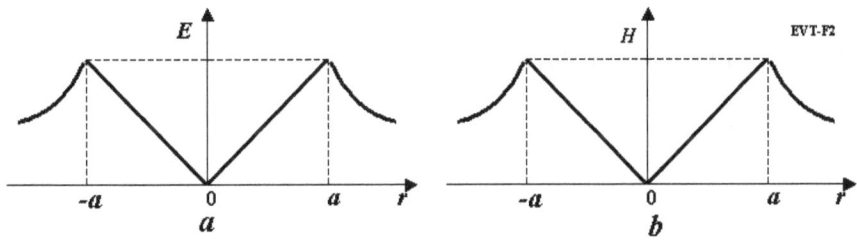

Fig. 4. Electrostatic and magnetic intensity into AB-Tube. *a*) Electrostatic intensity (pressure) via tube radius. *b*) Magnetic intensity (pressure) from electric current versus rube radius.

Theory of Plasma Transfer for Electric Energy, Estimations and Computations
Long Distance Wireless Transfer of Electricity on Earth.

The transferring of electric energy from one continent to other continent through ionosphere and the

Earth surface is described again. For this transferring we need two space towers of 100 km height, the towers must have a big conducting ball at their top end and underground (better, underwater) plates for decreasing the contact electric resistance (a good Earth ground). The contacting ball is a large (up to 100 - 200 m diameter) inflatable gas balloon having a conductivity layer (covering, or coating).

Let us to offer the method which allows computation of the parameters and possibilities of this electric line.

The electric resistance and other values for a conductive medium can be estimated by the equations:

$$R = \frac{U}{I} = \frac{1}{2\pi a \lambda}, \quad W = IU = 2\pi a \lambda U^2, \quad E_a = \frac{U}{2a}, \tag{1}$$

where R is the electric resistance of a conductive medium, Ω (for sea water $\rho = 0.3$ Ω·m); a is the radius of the contacting (source and receiving sphere) balloon, m; λ is the electric conductivity, (Ω·m)$^{-1}$; E_a is electric intensity on the balloon surface, V/m.

The conductivity λ of the *E*-layer of Earth's ionosphere as a rare ionized gas can be estimated by the equations:

$$\lambda = \frac{ne^2 \tau}{m_e}, \quad \text{where} \quad \tau = \frac{L}{v}, \quad L = \frac{kT_k}{\sqrt{2}\pi r_m^2 p}, \quad v^2 = \frac{8kT_k}{\pi m_e}, \tag{2}$$

where $n = 3.1 \times 10^9 \div 5 \times 10^{11}$ 1/m³ is density of free electrons in *E*-layer of Earth's ionosphere, 1/m³; τ is the time of electrons on their track, s; L is the length traversed by electrons on their track, m; v is the average electron velocity, in m/s; $r_m = 3.7 \times 10^{-10}$ (for hydrogen N$_2$) is diameter of gas molecule, m; $p = 3.2 \times 10^{-3}$ N/m² is gas pressure for altitude 100 km, N/m²; $m_e = 9.11 \times 10^{-31}$ is mass of electrons, kg.

The transfer power and efficiency are

$$W = IU, \quad \eta = 1 - R_c / R, \tag{3}$$

where R_c is common electric resistance of conductivity medium, Ω; R is total resistance of the electric system, Ω.

See the detailed computations in the Macro-Projects section.

Earth's ionosphere as the gigantic storage of electric energy. The Earth surface and Earth's ionosphere is gigantic spherical condenser. The electric capacitance and electric energy storied in this condenser can be estimated by equations:

$$C = \frac{4\pi\varepsilon_0}{1/R_0 - 1/(R_0 + H)} \approx 4\pi\varepsilon_0 \frac{R_0^2}{H}, \quad E = \frac{CU^2}{2}, \tag{4}$$

where C is capacity of condenser, C; $R_0 = 6.369 \times 10^6$ m is radius of Earth; H is altitude of *E*-layer, m; $\varepsilon_0 = 8.85 \times 10^{-12}$ F/m is electrostatic constant; E is electric energy, J.

The leakage currency is

$$i = \frac{3\pi\lambda_a R_0^2}{H} U, \quad \lambda_a = n_a e \mu, \quad R_a = \frac{H}{4\pi\lambda_a R_0^2}, \quad t = CR_a, \qquad (5)$$

where i leakage currency, A; λ_a is conductivity of Earth atmosphere, $(\Omega \cdot m)^{-1}$, n_a is free electron density of atmosphere, $1/m^3$; $\mu = 1.3 \times 10^{-4}$ (for N_2) is ion mobility, $m^2/(sV)$; R_a is Earth's atmosphere resistance, Ω; t is time of discharging in $e = 2.73$ times, s.

Theory and Computation of Electronic Tube

Below the interested reader may find the evidence of main equations, estimations, and computations.

1. Relation between the linear electric charge of tube and electron gas pressure on tube surface:

$$p = \frac{\varepsilon_0 E^2}{2}, \quad E = k\frac{2\tau}{r}, \quad \varepsilon_0 = \frac{1}{4\pi k}, \quad \tau = \sqrt{\frac{2\pi r p}{k}}, \qquad (6)$$

where p is electron pressure, N/m^2; $\varepsilon_0 = 8.85 \times 10^{-12}$ F/m –electrostatic constant; $k = 9 \times 10^9$ Nm^2/C^2 is electrostatic constant; E is electric intensity, V/m; τ is linear charges of tube, C/m; r is radius of tube, m.

Example, for atmospheric pressure $p = 10^5$ N/m^2 we receive $E = 1.5 \times 10^8$ V/m, N/C, the linear charge $\tau = 0.00833r$ C/m.

2. Density of electron (ion) in 1 m^3 in tube.

$$n = \frac{\tau}{\pi r^2 e} = \frac{1}{2\pi e k}\frac{E}{r} = 1.1 \cdot 10^8 \frac{E}{r},$$
$$M_e = m_e n, \quad M_i = \mu n_p n, \quad \mu = \frac{m_i}{m_p}, \qquad (7)$$

where n is charge (electron or ion) density, $1/m^3$; $e = 1.6 \times 10^{-19}$ C is charge of electron; $m_e = 9.11 \times 10^{-31}$ is mass of electron, kg; $m_p = 1.67 \times 10^{-27}$ is mass of proton, kg; M_e is mass density of electron, kg/m^3; M_i is mass density of ion, kg/m^3.

For electron pressure 1 atm the electron density (number particles in m^3) is $n = 1.65 \times 10^{16}/r$.

3. Electric resistance of AD-tube. We estimate the friction of electron about the tube wall by gas-kinetic theory

$$F_B = \eta_B SV, \quad \eta_B = \frac{1}{6}\rho V, \quad \rho = m_e n,$$
$$\overline{F} = \frac{F}{S} = \frac{1}{6} m_e n V^2, \quad V = \frac{j}{en} = \frac{i}{en\pi r^2}, \qquad (8)$$

where F_B is electron friction, N; η_B is coefficient of friction; S is friction area, m^2; V is electron speed, m/s; ρ is density of electron gas, kg/m^3; $\overline{F}$ is relative electron friction, N/m^2; j is current density, A/m^2.

4. Electric loss. The electric loss (power) into tube is

$$P_T = \overline{F}_B SV, \quad S = 2\pi rL, \quad P_T = \frac{1}{3}\pi m_e nrLV^3,$$

$$P_T = \frac{m_e}{3e^3\pi^2}\frac{i^3 L}{n^2 r^5} = 7.5\cdot 10^{24}\frac{i^3 L}{n^2 r^5} \quad [W], \tag{9}$$

where P_T is electric loss, W; L is tube length, m; i is electric current, A.

5. Relative electric loss is

$$\overline{P}_T = \frac{P_T}{P}, \quad P = iU, \quad \overline{P}_T = \frac{m_e}{3\pi^2 e^3}\frac{i^2}{n^2 r^5}\frac{L}{U} = 7.5\cdot 10^{24}\frac{i^2}{n^2 r^5}\frac{L}{U} = 7.4\cdot 10^{25}\frac{j^2}{n^2 r}\frac{L}{U}, \tag{10}$$

Compare the relative loss the offered electric (tube) line and conventional electric long distance line. Assume the electric line have length $L = 2000$ km, electric voltage $U = 10^6$ V, electric current $i = 300$ A, atmospheric pressure into tube. For offered line having tube $r = 1$ m the relative loss equals $\overline{P}_T = 0.005$. For conventional electric line having cross section copper wire 1 cm² the relative loss is $\overline{P}_T = 0.105$. That is in 21 times more than the offered electric line. The computation of Equation (10) for atmospheric pressure and for ratio $L/U = 1$ are presented in fig. 5. As you see for electric line $L = 1000$ km, voltage $U = 1$ million V, tube radius $r = 2.2$ m, the electric current $i = 50$ A, the relative loss of electric power is one/millionth (10^{-6}), (only 50 W for transmitted power 50 millions watt!). For connection Earth's surface with ionosphere we need only 100 km electronic tube ir 100 km electrostatic tower [6].

Moreover, the offered electric line is cheaper by many times, may be levitated into the atmosphere at high altitude, does not need a mast and ground, doesn't require expensive copper, does not allow easy surface access to line tapping thieves who wish to steal the electric energy. And this levitating electric line may be suspended with equal ease over sea as over land.

6. Lift force of tube ($L_{F,1}$, kg/m) and mass of 1 m length of tube (W_1, kg/m) is

$$L_{F,1} = \rho v = \rho\pi r^2; \quad W_1 = 2\pi r^2\gamma\delta, \tag{11}$$

where ρ is air density, at sea level $\rho = 1.225$ kg/m³; v is volume of 1 m of tube length, m³; γ is density of tube envelope, for most plastic $\gamma = 1500 \div 1800$ kg/m³; δ is film thickness, m.

Example. For $r = 10$ m and $\delta = 0.1$ mm, the lift force is 384 kg/m and cover mass is 11.3 kg/m.

7. Artificial fiber and tube (cable) properties [16]-[19]. Cheap artificial fibers are currently being manufactured, which have tensile strengths of 3-5 times more than steel and densities 4-5 times less than steel. There are also experimental fibers (whiskers) that have tensile strengths 30-100 times more than steel and densities 2 to 5 times less than steel. For example, in the book [16] p.158 (1989), there is a fiber (whisker) C_D, which has a tensile strength of $\sigma = 8000$ kg/mm² and density (specific gravity) of $\gamma = 3.5$ g/cm³. If we use an estimated strength of 3500 kg/mm² ($\sigma = 7\cdot 10^{10}$ N/m², $\gamma = 3500$ kg/m³), than the ratio is $\gamma/\sigma = 0.1\times 10^{-6}$ or $\sigma/\gamma = 10\times 10^6$.

Although the described (1989) graphite fibers are strong ($\sigma/\gamma = 10\times 10^6$), they are at least still ten times weaker than theory predicts. A steel fiber has a tensile strength of 5000 MPA (500 kg/sq.mm), the theoretical limit is 22,000 MPA (2200 kg/mm²) (1987); polyethylene fiber has a tensile strength

20,000 MPA with a theoretical limit of 35,000 MPA (1987). The very high tensile strength is due to its nanotube structure [18].

Apart from unique electronic properties, the mechanical behavior of nanotubes also has provided interest because nanotubes are seen as the ultimate carbon fiber, which can be used as reinforcements in advanced composite technology. Early theoretical work and recent experiments on individual nanotubes (mostly MWNT's, Multi Wall Nano Tubes) have confirmed that nanotubes are one of the stiffest materials ever made. Whereas carbon-carbon covalent bonds are one of the strongest in nature, a structure based on a perfect arrangement of these bonds oriented along the axis of nanotubes would produce an exceedingly strong material. Traditional carbon fibers show high strength and stiffness, but fall far short of the theoretical, in-plane strength of graphite layers by an order of magnitude. Nanotubes come close to being the best fiber that can be made from graphite.

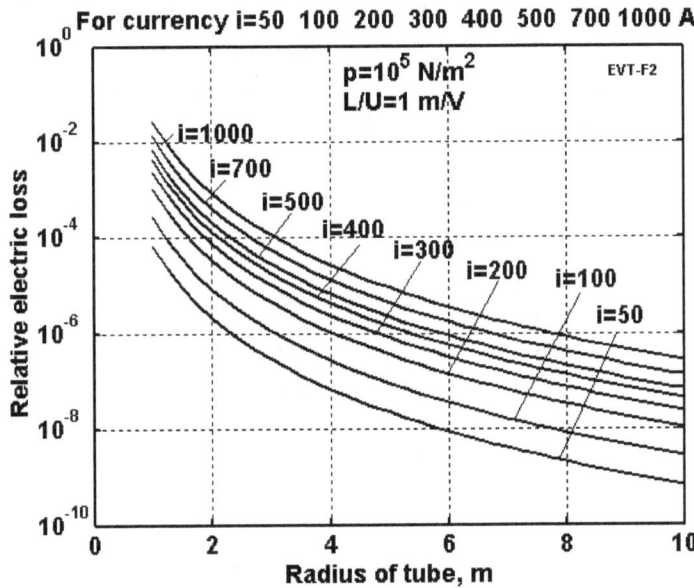

Fig. 5. Relative electric loss via radius of tube for electric **current** $i = 50 \div 1000$ A, the atmospheric pressure into tube and ratio $L/U = 1$.

For example, whiskers of Carbon nanotube (CNT) material have a tensile strength of 200 Giga-Pascals and a Young's modulus over 1 Tera Pascals (1999). The theory predicts 1 Tera Pascals and a Young's modules of 1-5 Tera Pascals. The hollow structure of nanotubes makes them very light (the specific density varies from 0.8 g/cc for SWNT's (Single Wall Nano Tubes) up to 1.8 g/cc for MWNT's, compared to 2.26 g/cc for graphite or 7.8 g/cc for steel). Tensile strength of MWNT's nanotubes reaches 150 GPa.

In 2000, a multi-walled carbon nanotube was tested to have a tensile strength of 63 GPa. Since carbon nanotubes have a low density for a solid of 1.3-1.4 g/cm³, its specific strength of up to 48,000 kN·m/kg is the best of known materials, compared to high-carbon steel's 154 kN·m/kg.

The theory predicts the tensile stress of different types of nanotubes as: Armchair SWNT - 120 GPa, Zigzag SWNT – 94 GPa.

Specific strength (strength/density) is important in the design of the systems presented in this paper; nanotubes have values at least 2 orders of magnitude greater than steel. Traditional carbon fibers have a specific strength 40 times that of steel. Since nanotubes are made of graphitic carbon, they have good resistance to chemical attack and have high thermal stability. Oxidation studies have shown that the onset of oxidation shifts by about 100^0 C or higher in nanotubes compared to high modulus graphite fibers. In a vacuum, or reducing atmosphere, nanotube structures will be stable to any practical service temperature (in vacuum up 2800 °C. in air up 750°C).

In theory, metallic nanotubes can have an electric current density (along axis) more than 1,000 times greater than metals such as silver and copper. Nanotubes have excellent heat conductivity along axis up 6000 W/m·K. Copper, by contrast, has only 385 W/m·K.

About 60 tons/year of nanotubes are produced now (2007). Price is about $100 - 50,000/kg. Experts predict production of nanotubes on the order of 6000 tons/year and with a price of $1 – 100/kg to 2012.

Commercial artificial fibers are cheap and widely used in tires and countless other applications. The authors have found only older information about textile fiber for inflatable structures (Harris J.T., Advanced Material and Assembly Methods for Inflatable Structures, AIAA, Paper No. 73-448, 1973). This refers to DuPont textile Fiber **B** and Fiber **PRD-49** for tire cord. They are 6 times strong as steel (psi is 400,000 or 312 kg/mm²) with a specific gravity of only 1.5. Minimum available yarn size (denier) is 200, tensile module is 8.8×10^6 (**B**) and 20×10^6 (**PRD-49**), and ultimate elongation (percent) is 4 (**B**) and 1.9 (**PRD-49**). Some data are in Table 1.

Table 1. Material properties

Material Whiskers	Tensile strength kg/mm²	Density g/cm³	Fibers	Tensile strength kg/mm²	Density g/cm³
AlB_{12}	2650	2.6	QC-8805	620	1.95
B	2500	2.3	TM9	600	1.79
B_4C	2800	2.5	Allien 1	580	1.56
TiB_2	3370	4.5	Allien 2	300	0.97
SiC	1380-4140	3.22	Kevlar or Twaron	362	1.44
Material			Dynecta or Spectra	230-350	0.97
Steel prestressing strands	186	7.8	Vectran	283-334	0.97
Steel Piano wire	220-248		E-Glass	347	2.57
Steel A514	76	7.8	S-Glass	471	2.48
Aluminum alloy	45.5	2.7	Basalt fiber	484	2.7
Titanium alloy	90	4.51	Carbon fiber	565	1,75
Polypropylene	2-8	0.91	Carbon nanotubes	6200	1.34

Source: [16]-[19] and Howatsom A.N., Engineering Tables and Data, p.41.

Industrial fibers have up to $\sigma = 500 - 600$ kg/mm², $\gamma = 1500 - 1800$ kg/m³, and $\sigma/\gamma = 2,78 \times 10^6$. But we are projecting use in the present projects the cheapest films and cables applicable (safety $\sigma = 100 - 200$ kg/mm²).

8. Dielectric strength of insulator. As you see above, the tube needs film that separates the positive charges located in conductive layer from the electron gas located in the tube. This film must have a high dielectric strength. The current material can keep a high E (see table 2 is taken from [10]).

Table 2. Properties of various good insulators (recalculated in metric system)

Insulator	Resistivity Ohm-m	Dielectric strength, MV/m. E_i	Dielectric con-stant, ε

Material			
Lexan	10^{17}–10^{19}	320–640	3
Kapton H	10^{19}–10^{20}	120–320	3
Kel-F	10^{17}–10^{19}	80–240	2–3
Mylar	10^{15}–10^{16}	160–640	3
Parylene	10^{17}–10^{20}	240–400	2–3
Polyethylene	10^{18}–5×10^{18}	40–680*	2
Poly (tetra-fluoraethylene)	10^{15}–5×10^{19}	40–280**	2
Air (1 atm, 1 mm gap)		4	1
Vacuum (1.3×10^{-3} Pa, 1 mm gap)		80–120	1

*For room temperature 500 – 700 MV/m.
** 400–500 MV/m.

Sources: Encyclopedia of Science & Technology (New York, 2002, Vol. 6, p. 104, p. 229, p. 231) and Kikoin [17] p. 321.

Note: Dielectric constant ε can reach 4.5 - 7.5 for mica (E is up 200 MV/m), 6 -10 for glasses ($E = 40$ MV/m), and 900 - 3000 for special ceramics (marks are CM-1, T-900) [17], p. 321, ($E = 13 - 28$ MV/m). Ferroelectrics have ε up to 10^4 - 10^5. Dielectric strength appreciably depends from surface roughness, thickness, purity, temperature and other conditions of materials. Very clean material without admixture (for example, quartz) can have electric strength up 1000 MV/m. As you see, we have the needed dielectric material, but it is necessary to find good (and strong) isolative materials and to research conditions which increase the dielectric strength.

9. Tube cover thickness. The thickness of the tube's cover may be found from Equation

$$\delta = \frac{rp}{\sigma}, \tag{12}$$

where p is electron pressure minus atmospheric pressure, N/m². If electron pressure is little more then the atmospheric pressure the tube cover thickness may be very thin.

10. Mass of tube cover. The mass of tube cover is

$$M_1 = \delta \gamma, \quad M = 2\pi r L \gamma \delta, \tag{13}$$

where M_1 is 1 m² cover mass, kg/m²; M is cover mass, kg.

11. The volume V and surface of tube s are

$$V = \pi r^2 L, \quad s = 2\pi r L, \tag{14}$$

where V is tube volume, m³; s is tube surface, m².

12. Relation between tube volume charge and tube liner charge for neutral tube is

$$E_V = \frac{\rho r}{2\varepsilon_0}, \quad E_s = \frac{\tau}{2\pi\varepsilon_0 r}, \quad E_V = E_s, \quad \tau = \pi\rho r^2, \quad \rho = \frac{\tau}{\pi r^2}, \quad (15)$$

where ρ is tube volume charge, C/m³; τ is tube linear charge, C/m.

13. General charge of tube. We got equation from

$$\tau = 2\pi\varepsilon\varepsilon_0 Er, \quad Q = \tau L, \quad Q = 2\pi\varepsilon\varepsilon_0 ErL, \quad (16)$$

where Q is total tube charge, C; ε is dielectric constant (see Table 2).

14. Charging energy. The charged energy is computed by equation

$$W = 0.5QU, \quad U = \delta E, \quad W = 0.5Q\delta E, \quad (17)$$

where W is charge energy, J; U is voltage, V.

15. Mass of electron gas. The mass of electron gas is

$$M_e = m_e N = m_e \frac{Q}{e}, \quad (18)$$

where M_e is mass of electron gas, kg; $m_e = 9.11 \times 10^{-31}$ kg is mass of electron; N is number of electrons, $e = 1.6 \times 10^{-19}$ is the electron charge, C.

16. Transfer of matter (Matter flow of ion gas). If we change the electron gas by the ion gas, our tube transfer charged matter with very high speed

$$M = M_i \pi r^2 V, \quad M_i = \mu m_p n,$$
$$V = \frac{i}{en\pi r^2}, \quad M = \frac{m_p}{e}\mu = 1.04 \cdot 10^{-8} \mu i \quad (19)$$

where M is the mass flow, kg/s; M_i is the gas ion density, kg/m³; $\mu = m_i/m_p$; V is ions speed, m/s.

Example: We want to transfer to a remote location the nuclear breeder fuel – Uranium-238. ($\mu = 238$) by line having $i = 1000$ A, $r = 1$ m, ion gas pressure 1 atm. One day contains 86400 seconds.

The equation (19) gives $M = 214$ kg/day, speed $V = 120$ km/s. The AB-tubes are suitable for transferring small amounts of a given matter. For transferring a large mass the diameter of tube and electric current must be larger.

We must also have efficient devices for ionization and utilization of the de-ionization (recombination) energy.

The offered method allows direct conversion of the ionization energy of the electron gas or ion gas to light (for example, by connection between the electron and ion gases).

17. Electron gas pressure. The electron gas pressure may be computed by equation (11). This computation is presented in fig. 6.

As you see the electron pressure reaches 1 atm for an electric intensity 150 MV/m and for negligibly small mass of the electron gas.

18. Power for support of charge. Leakage current (power) through the cover may be estimated by equation

$$I = \frac{U}{R}, \quad U = \delta E = \frac{r\varepsilon_0 E}{\sigma}, \quad R = \rho\frac{\delta}{s}, \quad I = \frac{sE}{\rho}, \quad W_l = IU = \frac{\delta s E^2}{\rho}, \quad (20)$$

where I is electric **current**, A; U is voltage, V; R is electric resistance, Ohm; ρ is specific resistance, Ohm·m; s is tube surface area, m².

The estimation gives the support power has small value.

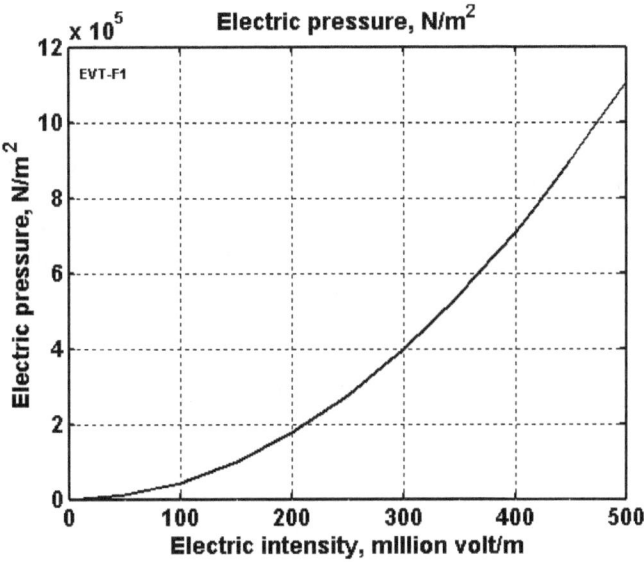

Fig. 6. Electron pressure versus electric intensity

Quasi-superconductivity of AB-Tube.
The proposed AB-Tube may become what we may term 'quasi-superconductive' when magnetic pressure equals electrostatic pressure. In this case electrons cannot contact with the tube wall, do not experience resistance friction and the AB-Tube thus experiences this 'quasi-superconductivity'.
Let us to get this condition:

$$P_e = \frac{\varepsilon_0 E^2}{2}, \quad P_m = \frac{B^2}{2\mu_0}, \quad P_e = P_m, \quad c = \frac{1}{\sqrt{\mu_0 \varepsilon_0}}, \quad E = cB, \quad (21)$$

where P_e is electronic pressure, N/m²; P_m is magnetic pressure, N/m²; B is magnetic intensity, T; E is electric intensity, V/m; c is light speed, $c = 3 \times 10^8$ m/s; ε_0, $\mu_0 = 4\pi \times 10^{-7}$ are electrostatic and magnetic constants. The relation $E = cB$ is important result and condition of tube superconductivity. For electron pressure into tube 1 atm, the $E = 1.5 \times 10^8$ V/m (see above) and $B = 0.5$ T.

From Eq. (21) we receive the relation between the electric current and the tube charge for AB-Tube 'quasi-superconductivity' as

$$E = cB, \quad E = \frac{1}{2\pi\varepsilon_0}\frac{\tau}{r}, \quad B = \frac{\mu_0 i}{2\pi r}, \quad i = c\tau, \qquad (22)$$

where i is electric current, A; τ is liner charge of tube, C/m.

For electron pressure equals 1 atm and $r = 1$m the linear tube charge is $\tau = 0.00833$ C/m (see above) and the request electric current is $i = 2.5 \times 10^6$ A ($j = 0.8$ A/m^2). For $r = 0.1$ m the current equals $i = 2.5 \times 10^5$ A. And for r = 0.01 m the current equals $i = 2.5 \times 10^4$ A.

Unfortunately, the requested electron speed (for true and full normal temperature 'superconductivity') equals light speed c.

$$V = \frac{j}{en} = \frac{i}{en\pi r^2} = \frac{c\tau}{en\pi r^2} = \frac{c\tau}{\tau} = c, \qquad (23)$$

That means we cannot exactly reach it, but we can came very close and we can have very low electric resistance of AB-Tube.

Information about high speed of electron and ion beam. Here $\gamma = (1 - \beta^2)^{-1/2}$ is the relativistic scaling factor, $\beta = v/c$, v is relative system speed; quantities in analytic formulas are expressed in SI or cgs units, as indicated; in numerical formulas I is in amperes (A), B is in gauss (G, 1 T = 10^4 G), electron linear density N is in cm^{-1}, temperature, voltage, and energy are in MeV, $\beta_z = v_z/c$, and k is Boltzmann's constant.

If the system is moved only along axis x, the Lorentz transformation are (" ' " is marked mobile system):

$$t' = \gamma\left(t - \frac{vx}{c^2}\right), \quad x' = \gamma(x - vt), \quad y' = y, \quad z' = z,$$

$$w' = \frac{w - v}{1 - wv/c^2}, \quad M = \gamma m, \quad \vec{p} = M\vec{v}, \quad \vec{f} = \frac{d\vec{p}}{dt}, \qquad (24)$$

where t is time, s; w is speed into systems, m/s, v is system speed, m/s, M is relativistic mass, kg; p is momentum, f is force, N.

For computation electrostatic and magnetic fields about light speed are useful the equations of relativistic theory (Lorenz's Equations, In the immobile system (market "$_1$")) the electric field is directed along axis y, the magnetic field is directed along axis z)):

$$E_x = E_{1x}, \qquad\qquad H_x = H_{1x},$$
$$\sqrt{1-\beta^2}\,E_y = E_{1y} - vB_{1z}, \quad \sqrt{1-\beta^2}\,H_y = H_{1y} + vD_{1z}, \qquad (25)$$
$$\sqrt{1-\beta^2}\,E_z = E_{1z} + vB_{1y}, \quad \sqrt{1-\beta^2}\,H_z = H_{1z} - vD_{1y},$$

where lower index "₁" means the immobile system coordinate, E is electric intensity, V/m; H is magnetic intensity, A/m; v is speed of mobile system coordinate along axis x, m/s; D is electric displacement. C/m²; $\beta = v/c$ is relative speed one system about the other.

Relativistic electron gyroradius [22]:

$$r_e = \frac{mc^2}{eB}(\gamma^2 - 1)^{1/2} \text{ (cgs)} = 1.70 \cdot 10^3 (\gamma^2 - 1)^{1/2} B^{-1} \quad \text{cm}. \tag{26}$$

Relativistic electron energy:

$$W = mc^2 \gamma = 0.511 \gamma \quad \text{MeV}. \tag{27}$$

Bennett pinch condition:

$$I^2 = 2Nk(T_e + T_i)c^2 \text{ (cgs)} = 3.20 \cdot 10^{-4} N(T_e + T_i) \quad \text{A}^2. \tag{28}$$

Alfven-Lawson limit:

$$I_A = (mc^3/e)\beta_z \gamma \text{ (cgs)} = (4\pi mc/\mu_0 e)\beta_z \gamma \text{ (SI)} = 1.70 \cdot 10^4 \beta_z \gamma \quad \text{A}. \tag{29}$$

The ratio of net current to I_A is

$$\frac{I}{I_A} = \frac{\nu}{\gamma}. \tag{30}$$

Here $\nu = Nr_e$ is the Budker number, where $r_e = e^2/mc^2 = 2.82 \cdot 10^{-13}$ cm is the classical electron radius. Beam electron number density is

$$n_b = 2.08 \cdot 10^8 J \beta^{-1} \quad \text{cm}^{-3}, \tag{31}$$

where J is the current density in A cm⁻². For a uniform beam of radius a (in cm):

$$n_b = 6.63 \cdot 10^7 I a^{-2} \beta^{-1} \quad \text{cm}^{-3} \tag{32}$$

and

$$\frac{2r_e}{a} = \frac{\nu}{\gamma}, \tag{33}$$

Child's law: nonrelativistic space-charge-limited current density between parallel plates with voltage drop V (in MV) and separation d (in cm) is

$$J = 2.34 \cdot 10^3 V^{3/2} d^{-2} \quad \text{A cm}^{-2} \tag{34}$$

The condition for a longitudinal magnetic field B_z to suppress filamentation in a beam of current density J (in A cm⁻²) is

$$B_z > 47 \beta_z (\gamma J)^{1/2} \quad \text{G}. \tag{35}$$

Kinetic energy necessary to accelerate a particle is

$$K = (\gamma - 1)mc^2. \qquad (36)$$

The de Broglie wavelength of particle is $\lambda = h/p$, where $h = 6.6262 \times 10^{-34}$ J·s is Planck constant, p is particle momentum. Classical radius of electron is 2.8179×10^{-15} m.

Macroprojects

Wireless transferring energy between Earth's continents (Fig. 2). Let us take the following initial data: Gas pressure at altitude 100 km is $p = 3.2 \times 10^{-3}$ N/m^2, temperature is 209 K, diameter nitrogen N_2 molecule is 3.7×10^{-10} m, the ion/electron density in ionosphere is $n = 10^{10}$ 1/m^3, radius of the conductivity inflatable balloon at top the space tower (mast) is $a = 100$ m (contact area is $S = 1.3 \times 10^5$ m^2), specific electric resistance of a sea water is 0.3 Ω·m, area of the contact sea plate is 1.3×10^3 m^2.

The computation used equation (1)-(2) and (15)-(17) [4] gives: electron track in ionosphere is $L = 1.5$ m, electron velocity $\upsilon = 9 \times 10^4$ m/s, track time $\tau = 1.67 \times 10^{-5}$ s, specific resistance of ionosphere is $\rho = 4.68 \times 10^{-3}$ $(\Omega \cdot m)^{-1}$, contact resistance of top ball (balloon) is $R_1 = 0.34$ Ω, contact resistance of the lower sea plates is $R_2 = 4.8 \times 10^{-3}$ Ω, electric intensity on ball surface is 5×10^4 V/m.

If the voltage is $U = 10^7$ V, total resistance of electric system is $R = 100$ Ω, then electric currency is $I = 10^5$ A, transferring power is $W = IU = 10^{12}$ W, coefficient efficiency is 99.66%. That is power 1000 powerful electric plants, having power one billion watts. In practice we are not limited in transferring any energy in any Earth's point having the 100 km space mast and further transfer by ground-based electric lines in any geographical region of radius 1000 ÷ 2000 km.

Earth's ionosphere as the storage electric energy. It is using the equations (18)-(19) [4] we find the Earth's-ionosphere capacity $C = 4.5 \times 10^{-2}$ C. If $U = 10^8$ V, the storage energy is $E = 0.5CU^2 = 2.25 \times 10^{14}$ J. That is large energy. About 20 of 100 tons rocket may be launched to space in 100 km orbit. This energy are produced a powerful electric plant in one day.

Let us now estimate the leakage of current. Cosmic rays and Earth's radioactivity create 1.5 ÷ 10.4 ions every second in 1 cm^3. But they quickly recombine in neutral molecule and the ions concentration is small. We take the ion concentration of lower atmosphere $n = 10^6$ 1/m^3. Then the specific conductivity of Earth's atmosphere is 2.1×10^{-17} $(\Omega.m)^{-1}$. The leakage currency is $i = 10^{-7} \times U$. The altitude of E-layer is 100 km. We take a thickness of atmosphere only 10 km. Then the conductivity of Earth's atmosphere is 10^{-24} $(\Omega.m)^{-1}$, resistance is $R_a = 10^{24}$ Ω, the leakage time (decreasing of energy in $e = 2.73$ times) is 1.5×10^5 years.

As you can clearly see the Earth's ionosphere may become a gigantic storage site of electricity.

The electric resistance of electronic tube is small.

Discussing

The offered ideas and innovations may create a jump in space and energy industries. Author has made initial base researches that conclusively show the big industrial possibilities offered by the methods and installations proposed.

The offered inflatable electrostatic AB tube has indisputably remarkable operational advantages in comparison with the conventional electric lines. AB-tube may be also used for transfer electricity in long distance without using ionosphere.

The main innovations and applications of AB-Tubes are:

1. Transferring electric energy in a long distance (up 10,000 km) with a small electric loss.
2. 'Quasi-superconductivity'. The offered AB-Tube may have a very low electric resistance for any temperature because the electrons in the tube do not have ions and do not lose energy by impacts with ions. The impact the electron to electron does not change the total impulse (momentum) of couple electrons and electron flow. If this idea is proved in experiment, that will be big breakthrough in many fields of technology.
3. Cheap electric lines suspended in high altitude (because the AB-Tube can have lift force in atmosphere and do not need ground mounted electric masts and other support structures)
4. The big diameter AB-Tubes (including the electric lines for internal power can be used as tramway for transportation .
5. AB-Tube s can be used as vacuum tubes for an exit from the Earth's surface to outer space (out from Earth's atmosphere). That may be used by an Earth telescope for observation of sky without atmosphere hindrances, or sending of a plasma beam to space ships without atmosphere hindrances [12-14].
6. Transfer of electric energy from continent to continent through the Earth's ionosphere [4-5].
7. Inserting an anti-gravitator cable into a vacuum-enclosing AB-Tube for near-complete elimination of air friction [4-5]. Same application for transmission of mechanical energy for long distances with minimum friction and losses. [4-5].
8. Increasing in some times the range of a conventional gun. They can shoot through the vacuum tube (up 4-6 km) and projectile will fly in the rare atmosphere where air drag is small.
9. Transfer of matter a long distance with high speed (including in outer space, see other of author's works).
10. Interesting uses in nuclear and high energy physics engineering (inventions).

The offered electronic gas may be used as filling gas for air balloons, dirigibles, energy storage, submarines, electricity-charge devices (see also [4]-[15]).

Further research and testing are necessary. As that is in science, the obstacles can slow, even stop, applications of these revolutionary innovations.

Summary

This new revolutionary idea - wireless transferring of electric energy in long distance through the ionosphere or by the electronic tubes is offered and researched. A rare plasma power cord as electric cable (wire) is used for it. It is shown that a certain minimal electric currency creates a compressed force that supports the plasma cable in the compacted form. Large amounts of energy can be transferred many thousands of kilometers by this method. The requisite mass of plasma cable is merely hundreds of grams. It is computed that the macroproject: The transfer of colossal energy from one continent to another continent (for example, Europe to USA and back), using the Earth's ionosphere as a gigantic storage of electric energy [1]-[21].

Acknowledgement

The author wishes to acknowledge R.B. Cathcart (USA) for helping to correct the author's English.

References

(Reader finds some of author's articles in http://Bolonkin.narod.ru/p65.htm , http://www.scribd.com and http://arxiv.org , search "Bolonkin", in books "*Non-Rocket Space Launch and Flight*", Elsevier, 2006, 488 pgs; "*New concepts, Ideas, and Innovation in Aerospace, Technology and Human Science*", NOVA, 2008, 502 pgs.; *Macro-Projects: Environment and Technology*, NOVA, 2009, 536 pgs.; "New Technologies and Revolutionary Projects", Sbcribd, 2010, 324 pgs, http://www.scribd.com/doc/32744477).

1. Bolonkin, A.A., (1982), Installation for Open Electrostatic Field, Russian patent application #3467270/21 116676, 9 July, 1982 (in Russian), Russian PTO.

2. Bolonkin, A.A., (1983), Method of stretching of thin film. Russian patent application #3646689/10 138085, 28 September 1983 (in Russian), Russian PTO.
3. Bolonkin, A.A., Getting of Electric Energy from Space and Installation for It, Russian patent application #3638699/25 126303, 19 August, 1983 (in Russian), Russian PTO.
4. Bolonkin A.A., Wireless Transfer of Electricity in Outer Space. Presented to http://Arxiv.org on 4 January, 2007. Presented as paper AIAA-2007-0590 to 45th AIAA Aerospace Science Meeting, 8 - 11 January 2007, Reno, Nevada, USA. http://aiaa.org search "Bolonkin" or http://Bolonkin.narod.ru/p65.htm .
5. Bolonkin A.A., AB Electronic Tubes and Quasi-Superconductivity at Room Temperature. Presented to http://arxiv.org on 8 April, 2008; http://Bolonkin.narod.ru/p65.htm .
6. 6. Bolonkin A.A., Optimal Electrostatic Space Tower , Presented as paper AIAA-2007-6201 to Space-2007 Conference, 18-20 September 2007, Long Beach, CA, USA. http://arxiv.org , search "Bolonkin", http://aiaa.org search "Bolonkin"; http://Bolonkin.narod.ru/p65.htm .
7. Bolonkin A.A., Optimal Solid Space Tower, AIAA-2006-7717. ATIO Conference, 25-27 Sept. 2006, Wichita, Kansas, USA. http://arxiv.org , http://aiaa.org search "Bolonkin"; http://Bolonkin.narod.ru/p65.htm .
8. Bolonkin A.A., Optimal Inflatable Space Tower with 3-100 km Height, Journal of the British Interplanetary Society, Vol.56, No. 3/4, 2003, pp.97-107. http://Bolonkin.narod.ru/p65.htm .
9. Bolonkin A.A., Kinetic Space Towers. Presented as paper IAC-02-IAA.1.3.03 at World Space Congress-2002 10-19 October, Houston, TX, USA. Detail Manuscript was published as A.A.Bolonkin. "Kunetif Space Towers and Launchers", Journal of the British Interplanetary Society, Vol.57, No.1/2, 2004. pp.33-39. http://Bolonkin.narod.ru/p65.htm .
10. Bolonkin A.A. Linear Electrostatic Engine, The work was presented as paper AIAA-2006-5229 for 42 Joint Propulsion Conference, Sacramento, USA, 9-12 July, 2005. Work is published in International journal AEAT, Vol.78, #6, 2006, pp.502-508. http://Bolonkin.narod.ru/p65.htm .
11. Bolonkin A.A., Krinker M., Magnetic Space Launcher, 45 Joint Propulsion Conference, USA, 2009. http://Bolonkin.narod.ru/p65.htm .
12. Bolonkin A.A., *Non-Rocket Space Launch and Flight*, Elsevier, London, 2006, 488 pgs. http://www.scribd.com/doc/24056182
13. Bolonkin A.A., "*New concepts, Ideas, and Innovation in Aerospace, Technology and Human Science*", NOVA, 2008, 502 pgs. http://www.scribd.com/doc/24057071 .
14. Bolonkin A.A., Cathcart R.B., "*Macro-Projects: Environment and Technology*", NOVA, 2009, 536 pgs. http://www.scribd.com/doc/24057930 .

15. *Macro-Engineering - A challenge for the future*. Collection of articles. Eds. V. Badescu, R. Cathcart and R. Schuiling, Springer, 2006. See article: "Space Towers" by A.Bolonkin..
16. Galasso F.S., Advanced Fibers and Composite, Gordon and Branch Science Publisher, 1989.
17. Kikoin, I.K., (ed.), Tables of physical values. Atomuzdat, Moscow, 1976 (in Russian).
18. Dresselhous M.S., Carbon Nanotubes, Springer, 2000.
19. AIP. Physics desk reference, 3-rd Edition, Springer, 2003.
20. Wikopedia. http://wikipedia.org . Space Towers.
21. Krinker M., Review of New Concepts, Ideas and Innovations in Space Towers. http://www.scribd.com/doc/26270139, http://arxiv.org/ftp/arxiv/papers/1002/1002.2405.pdf
22. Bolonkin A.A., Wireless transfer of electricity from continent to continent . http://www.scribd.com/doc/427211638 .

17 February 2009

Chapter 6

Non Turbo Electric Wind Generator

Abstract

Author offers a new method of getting electric energy from wind. A special injector injects electrons into the atmosphere. Wind picks up the electrons and moves them in the direction of wind which is also against the direction of electric field. At some distance from injector a unique grid acquires the electrons, thus charging and producing electricity. This method does not require, as does other wind energy devices, strong columns, wind turbines, or electric generators. This proposed wind installation is cheap. The area of wind braking may be large and produces a great deal of energy. Although this electron wind installations may be in a city, the population will not see them.

Keywords: *wind energy, utilization of wind energy, electronic wind electric generator, EABG, Bolonkin.*

Introduction

Wind power is the conversion of wind energy into a useful form of energy, such as using wind turbines to make electrical power, windmills for mechanical power, wind pumps for water pumping or drainage, or sails to propel ships.

Large wind farms consist of hundreds of individual wind turbines which are connected to the electric power transmission network. Offshore wind is steadier and stronger than on land, and offshore farms have less visual impact, but construction and maintenance costs are considerably higher. Small onshore wind farms provide electricity to isolated locations. Utility companies increasingly buy surplus electricity produced by small domestic wind turbines.

Wind power, as a viable alternative to fossil fuels, is plentiful, renewable, widely distributed, clean, produces no greenhouse gas emissions during operation and uses little land. The effects on the environment are generally less problematic than those from other power sources. As of 2011, Denmark generates more than a quarter of its electricity from wind and 83 countries around the world are using wind power on a commercial basis. In 2010 wind energy production was over 2.5% of total worldwide electricity usage, and growing rapidly at more than 25% per annum. The monetary cost per unit of energy produced is similar to the cost for new coal and natural gas installations.

Worldwide there are now over two hundred thousand wind turbines operating, with a total nameplate capacity of 282,482 MW as of end 2012. The European Union alone passed some 100,000 MW nameplate capacity in September 2012, while the United States surpassed 50,000 MW in August 2012 and China passed 50,000 MW the same month.

Some Information about Wind Energy. The power of wind engine strongly depends on wind speed (to the third power). Low altitude wind ($H = 10$ m) has the standard average speed of $V = 6$ m/s. High altitude wind is powerful and practically everywhere is stable and constant. Wind in the troposphere and stratosphere are powerful and permanent. For example, at an altitude of 5 km, the average wind speed is about 20 M/s, at an altitude 10 - 12 km the wind may reach 40 m/s (at latitude of about 20 - 35^0 N).

There are permanent jet streams at high altitude. For example, at $H = 12$-13 km and about 25^0 N latitude, the average wind speed at its core is about 148 km/h (41 m/s). The most intensive portion has a maximum speed of 185 km/h (51 m/s) latitude 22^0, and 151 km/h (42 m/s) at latitude 35^0 in North America. On a given winter day, speeds in the jet core may exceed 370 km/h (103 m/s) for a distance of several hundred miles along the direction of the wind. Lateral wind shears in the direction normal to the jet stream may be 185 km/h per 556 km to right and 185 km/h per 185 km to the left.

The wind speed of $V = 40$ m/s at an altitude $H = 13$ km provides 64 times more energy than surface wind speeds of 6 m/s at an altitude of 10 m. This is an enormous renewable and free energy source. (See reference: *Science and Technology, v.2,* p.265).

Economy of conventional utilization of wind energy. Current wind power plants have low ongoing costs, but moderate capital cost. The marginal cost of wind energy once a plant is constructed is usually less than 1-cent per kW·h. The estimated average cost per unit incorporates the cost of construction of the turbine and transmission facilities, borrowed funds, return to investors (including cost of risk), estimated annual production, and other components, averaged over the projected useful life of the equipment, which may be in excess of twenty years. Energy cost estimates are highly dependent on these assumptions so published cost figures can differ substantially. In 2004, conventional wind energy cost a fifth of what it did in the 1980s, and a continued downward trend is expected as larger multi-megawatt turbines were mass-produced. A 2011 report from the American Wind Energy Association stated, "Wind's costs have dropped over the past two years, in the range of 5 to 6 cents per kilowatt-hour recently.... about 2 cents cheaper than coal-fired electricity, and more projects were financed through debt arrangements than tax equity structures last year.... winning more mainstream acceptance from Wall Street's banks.... Equipment makers can also deliver products in the same year that they are ordered instead of waiting up to three years as was the case in previous cycles.... 5,600 MW of new installed capacity is under construction in the United States, more than double the number at this point in 2010. Thirty-five percent of all new power generation built in the United States since 2005 has come from wind, more than new gas and coal plants combined, as power providers are increasingly enticed to wind energy as a convenient hedge against unpredictable commodity price moves."

A British Wind Energy Association report gives an average generation cost of onshore wind power of around 3.2 pence (between US 5 and 6 cents) per kW·h (2005). Cost per unit of energy produced was estimated in 2006 to be comparable to the cost of new generating capacity in the US for coal and natural gas: wind cost was estimated at $55.80 per MW·h, coal at $53.10/MW·h and natural gas at $52.50. Similar comparative results with natural gas were obtained in a governmental study in the UK in 2011. A 2009 study on wind power in Spain by Gabriel Calzada Alvarez of King Juan Carlos University concluded that each installed MW of wind power led to the loss of 4.27 jobs, by raising energy costs and driving away electricity-intensive businesses. The U.S. Department of Energy found the study to be seriously flawed, and the conclusion unsupported. The presence of wind energy, even when subsidized, can reduce costs for consumers (€5 billion/yr in Germany) by reducing the marginal price, by minimizing the use of expensive peaking power plants.

In February 2013 Bloomberg New Energy Finance reported that the cost of generating electricity from new wind farms is cheaper than new coal or new baseload gas plants. In Australia, when including the current Australian federal government carbon pricing scheme their modeling gives costs (in Australian dollars) of $80/MWh for new wind farms, $143/MWh for new coal plants and $116/MWh for new baseload gas plants. The modeling also shows that "even without a carbon price (the most efficient way to reduce economy-wide emissions) wind energy is 14% cheaper than new coal and 18% cheaper than new gas." Part of the higher costs for new coal plants is due to high financial lending costs because of "the reputational damage of emissions-intensive investments". The expense of gas fired plants is partly due to "export market" effects on local prices. Costs of production from coal fired plants built in "the 1970s and 1980s" are cheaper than renewable energy sources because of depreciation.

Programs for Developing Wind Energy. Wind is a clean and inexhaustible source of energy that has been used for many centuries to grind grain, pump water, propel sailing ships, and perform other work. Wind farm is the term used for a large number of wind machines clustered at a site with persistent favorable winds, generally near mountain passes. Wind farms have been erected in New Hampshire, in the Tehachapi Mountains, at Altamont Pass in California, at various sites in Hawaii, and may other locations. Machine capacities range from 10 to 500 kilowatts. In 1984 the total energy output of all wind farms in the United States exceeded 150 million kilowatt-hours.

A program of the United States Department of Energy encouraged the development of new machines, the construction of wind farms, and an evaluation of the economic effect of large-scale use of wind power.

The utilization of renewable energy ('green' energy) is currently on the increase. For example, numerous wind turbines are being installed along the British coast. In addition, the British government has plans to develop off-shore wind farms along their coast in an attempt to increase the use of renewable energy sources. A total of $2.4 billion was injected into renewable energy projects over the last three years in an attempt to meet the government's target of using renewable energy to generate 10% of the country's energy needs by 2010. This British program saves the emission of almost a million tons of carbon dioxide. Denmark plans to get about 30% of their energy from wind sources.

Unfortunately, current ground wind energy systems have deficiencies which limit their commercial applications:

5. Wind energy is unevenly distributed and has relatively low energy density. Huge turbines cannot be placed on the ground; many small turbines must be used instead. In California, there are thousands of small wind turbines. However, while small turbines are relatively inefficient, very huge turbines placed at ground are also inefficient due to the relatively low wind energy density and their high cost. The current cost of wind energy is higher than energy of thermal power stations.

6. Wind power is a function of the cube of wind velocity. At surface level, wind has low speed and it is non-steady. If wind velocity decreases in half, the wind power decreases by a factor of 8 times.

7. The productivity of a wind-power system depends heavily on the prevailing weather.

8. Wind turbines produce noise and visually detract from the landscape.

While there are many research programs and proposals for wind driven power generation systems, all of them are ground or tower based. The system proposed in this article is located at high altitude (up to the stratosphere), where strong permanent and steady streams are located. This article also proposes a solution to the main technologist challenge of this system; the transfer of energy to the ground via a mechanical transmission made from closed loop, modern composite fiber cable.

The reader can find the information about this idea in [1]-[2], a detailed description of the innovation in [3]-[6], and the wind energy in references [7]-[8], new material used in the proposed innovation in [9]-[13]. The review of last airborne concepts in [14]-[17].

Description of Innovation

One simplest version of the offered electron wind generator (EABG) is presented in fig.1. Installation contains: electron injectors 2 established in column 6 and electron collector (net) 4 having the conductive leaves 5 (metallic foil, for example, aluminum foil). They have a large surface which helps to collect the electrons from big area. Network connects with the electron injectors through a useful load 7.

Work of EABG. The EABG generator works the following way: injector injects the electrons into air, the wind catch them and moves to collector (network) 4. Network 4 has negative charge, electron injector has positive charge. The electric field breaks the electrons (negative ions) and decreases the wind speed. But the electric ion speed is less than wind speed and electrons when they reach the collector settle into collector and increase its negative charge. Those additional charges (electrons) return through the electric load 7 and make the useful work.

In the city any building may be used as an electron collector (fig.2). This building must be colored by a conductive paint. This layer of paint must be isolated from the Earth and connected to the injectors via useful electric load. The injectors may be located in other buildings or any electric, lamp, or telephone posts.

The injectors are located around the building and get wind energy regardless the directions of wind.

In places where there are no buildings, the collector is located on the Earth surface (fig.3). The injectors may be up on a mast (fig. 3a) or located also on earth surface (fig. 3b). The efficiency of these will be different. The surface collector is conductivity film 11 (fig.3) (for example, aluminum foil), isolated from Earth. For increasing the efficiency of collector we can (optionally) place under collector the isolated positive charge 12 (or positive electrets) (fig. 3).

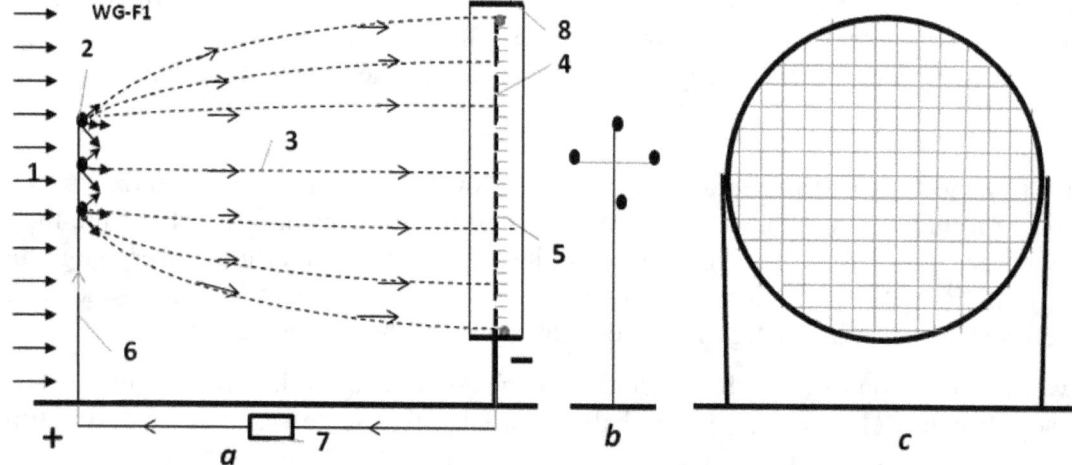

Fig.1. One version of Electron Wind Electric Generator (EABG). *a* – side view of the installation; *b* – front view of the electron injector column; c – front view of the collect net. *Notations:* 1 is wind; 2 is electron injector; 3 is trajectories of electrons; 4 is net collecting the electrons; 5 is conductive leaves (metallic foil, for example, aluminum foil); 6 is column (post) for supporting of the electron injectors; 7 is the outer electric load; 8 is high voltage ring of collector.

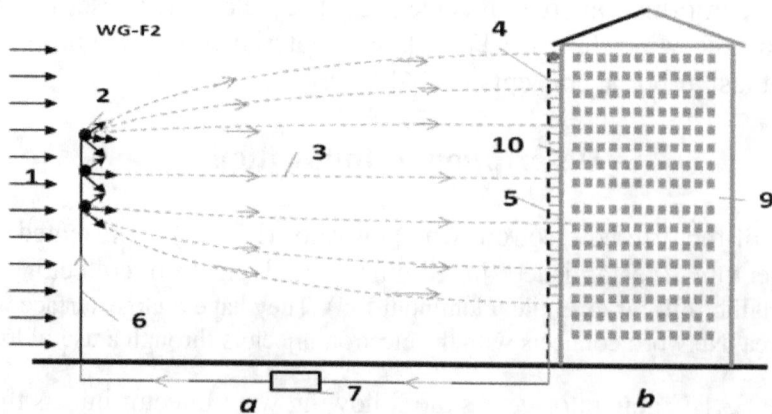

Fig.2. Using the building walls as the collector (net) for electrons.

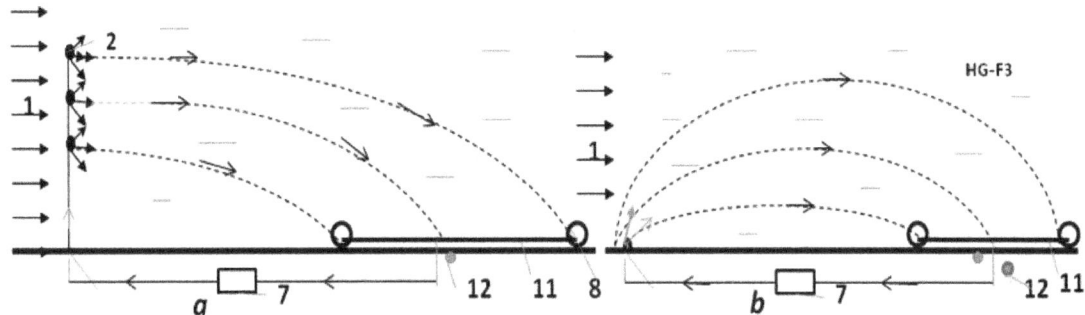

Fig.3. The horizontal conductivity film as collector of electrons. *a* – injectors in column; *b* - injectors at Earth surface. *Notations:* the number 1, 2 are same fig.1; 8 – ring of high voltage collector; 11 - conductivity film (for example, aluminum foil); 12 (optional) positive isolated charge (for example, electrets).

If we want to use wind energy at high altitudes, a special parachute can be used. Two versions of these designs are shown in fig.4. In the first version the electron injector is supported by wing 13 (fig.4a), in the second version (fig.4b) the electron injector is supported by a unique parachute 15 which creates also the lift force. Special parachute is net containing the conductive leaves as 5 in fig.1.

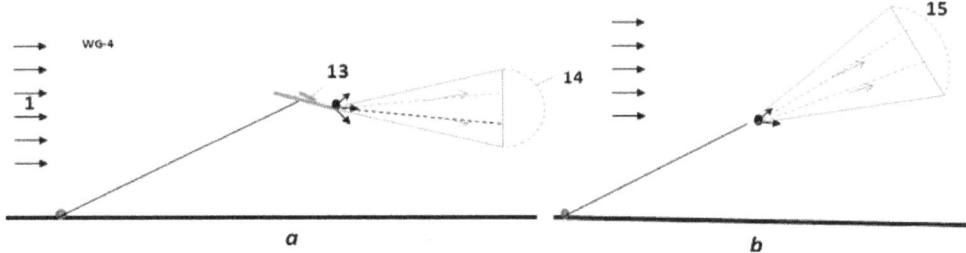

Fig.4. Airborne (flight) high altitude Electron wind generator. *a* - wing support; *b* – wind parachute support. *Notations:* 13 is wing; 14 is parachute; 15 is parachute having lift force.

Advantages of the proposed electron wind systems (EABG) in comparison with the conventional air wind systems.

The suggested new principle electron wind generator (EABG) has the following advantages in comparison with conventional wind systems used at present time.
Advantages:
1. Offered installations are very simple.
2. Offered system is very cheap (by hundreds of times). No tower, propeller, magnetic electric generator,
   gear box.
3. Offered system can cover a large area and has enormous power.
4. Offered installations are suitable for city having many high rise building.
5. The EABGs are invisible for population.
6. Offered installations produce high voltage direct electricity. That is advantage if energy is being
   transferred long distances.
7. Offered system is very suitable for airborne wind installation, because it is very light and produces high
   voltage electricity.
8. Offered system may be used as brake and can supply power to the electric system of aircraft.

Estimations and Computation

1. **Power of a wind** energy N [Watt, Joule/sec]

$$N = 0.5 \eta \rho A V^3 \quad [W] . \tag{1}$$

The coefficient of efficiency, η, equals about $0.2 \div 0.25$ for EABG; $0.15 \div 0.35$ for low speed propeller rotors (ratio of blade tip speed to wind speed equals $\lambda \approx 1$); $\eta = 0.45 \div 0.5$ for high speed propeller rotors ($\lambda = 5 - 7$). The Darrieus rotor has $\eta = 0.35 - 0.4$. The gyroplane rotor has $0.1 \div 0.15$. The air balloon and the drag (parachute) rotor has $\eta = 0.15 - 0.2$. The Makani rotor has $0.15 \div 0.25$. The theoretical maximum equals $\eta \approx 0.6$. Theoretical maximum of the electron generator is 0.25. A - front (forward) area of the electron corrector, rotor, air balloon or parachute [m^2]. ρ - density of air: ρ_o =1.225 kg/m^3 for air at sea level altitude $H = 0$; $\rho = 0.736$ at altitude $H = 5$ km; $\rho = 0.413$ at $H = 10$ km. V is average annually wind speed, m/s.

Table 1. Relative density ρ_r and temperature of the standard atmosphere via altitude

H, km	0	0.4	1	2	3	6	8	10	12
$\rho_r = \rho/\rho_o$	1	0.954	0.887	0.784	0.692	0.466	0.352	0.261	0.191
T, K	288	287	282	276	269	250	237	223	217

Issue [6].

The salient point here is that the strength of wind power depends upon the wind speed (by third order!). If the wind speed increases by two times, the power increases by 8 times. If the wind speed increases 3 times, the wind power increases 27 times!

The wind speed increases in altitude and can reach in constant air stream at altitude $H = 5 - 7$ km up $V = 30 - 40$ m/s. At altitude the wind is more stable/constant which is one of the major advantages that an airborne wind systems has over ground wind systems.

For comparison of different wind systems of the engineers must make computations for average annual wind speed $V_0 = 6$ m/s and altitude $H_0 = 10$ m. For standard wind speed and altitude the maximal wind power equals 66 W/m^2.

The energy, E, produced in one year is (1 year $\approx 30.2 \times 10^6$ work sec) [J]

$$E = 3600 \times 24 \times 350 N \approx 30 \times 10^6 N, \quad [J]. \tag{2}$$

2. **Electron speed**. The electron speed about the wind, gas (air) jet may be computed by equation:

$$j_s = q n_- b_- E + q D_-(dn_-/dx) , \tag{3}$$

where j_s is density of electric currency about jet, A/m^2; $q = 1.6 \times 10^{-19}$ C is charge of single electron, C; n_- is density of injected electrons (negative charges) in 1 m^3; b_- is charge mobility of negative charges, m^2/sV; E is electric intensity, V/m; D_- is diffusion coefficient of charges; dn_-/dx is gradient of charges. For our estimation we put $dn_-/dx = 0$. In this case

$$j_s = q n_- b_- E, \quad Q = q n, \quad v = b E, \quad j_s = Q v , \tag{4}$$

where Q is density of the negative charge in 1 m^3; v is speed of the negative charges about wind, m/s.

The negative charge mobility for normal pressure and temperature $T = 20°$C is:

In dry air $b_- = 1.9 \times 10^{-4}$ m²/sV, in humid air $b_- = 2.1 \times 10^{-4}$ m²/sV. (5)

If the air pressure is from 13 to 6×10^6 Pa, then the mobility follows the law bp = const, where p is air pressure. When air density decreases, the charge mobility increases. The mobility stregnth depends upon the purity of gas.

For normal air density the electric intensity must be less than 3 MV ($E <$ 3 MV). Otherwise the electric breakdown may be.

If $v > 0$, the electrons accelerate the air ($E > 0$ and installation spends energy, works as ventilator). If $v < 0$, the electrons beak the wind ($E < 0$ and the correct installation can produce energy, works as electric generator). If $v = 0$ (electron speed about installation equals wind speed V), the electric resistance is zero.

3. Optimal regime of work the electron generator. Let us to find the maximal power of electron generator.

The specific power of electron generator P [W/m²] is
$$P = Tv = 0.5\rho(V-v)^2 v , \quad (6)$$
where T is air trust, N/m²; V is wind speed, m/s; v is electron speed about air in opposed direction, m/s.

This function has maximum when relation
$$v/V = 1/3. \quad (7)$$

That means the optimal electric intensity is (see (2) – (3))($b = 2 \times 10^{-4}$):
$$v = bE, \quad E = v/b = V/3b = 1.67 \times 10^3 V, \quad [V/m] \quad (8)$$
where V is wind speed, m/s.

The optimal voltage and electric currency aproximatly is:
$$U \approx EL, \quad I = N/U, \quad (9)$$
where U is voltage, V; L is distance between injector and collector, m; I is electric currency, A.

4. Electron injectors.

There are some methods for generating electron emissions: hot cathode emission, cold field electron emission (edge cold emission, edge cathode), photo emission, radiation emission, radioisotope emission and so on. We consider only the hot emission and briefly the cold field electron emission (edge cathodes).

The **hot cathode** emission computed by equation:
$$j_s = BT^2 exp(-A/kT), \quad (10)$$
where B is coefficient, A/cm²K²; T is catode temperature, K; $k = 1.38 \times 10^{-23}$ [J/K] is Bolzmann constant; A is thermoelectron exit work, eV. Both values A, B depend from material of cathode and its cover. The "A" changes from 1.6 to 5 eV, the "B" changes from 0.5 to120 A/cm²K². Boron thermo-cathode produces electric currency up 200 A/cm². For temperature 1400 ÷1500K the cathode can produce currency up 1000 A/cm². The life of cathode can reach some years [19]-[20].

The edge cold emission. The cold field electron emission uses the edge cathodes. It is known that the electric intensity E_e in the edge is
$$E_e = U/a. \quad (11)$$
Here a is radious of the edge. If voltage between the edge and nears net (anode) is U = 1000 V, the

radius of edge $a = 10^{-5}$ m, electric intensity at edge is the $E_a = 10^8$ V/m. That is enough for the electron emission. The density of electric current may reach up 10^4 A/cm^2. For getting the required currency we make the need number of edges.

5. Airborne wind Turbine.

The drag of the vertical collector/rotor equals
$$D_r = N/V, \quad [N]. \tag{12}$$

The lift force of the wing, L_w, is
$$L_w = 0.5 C_L \rho V^2 A_w, \quad [N], \tag{13}$$
where C_L is lift coefficient (maximum $C_L \approx 2 - 2.5$); A_w is area of the wing, m^2.

The drag of the wing is
$$D_w = 0.5 C_D \rho V^2 A_w, \quad [N], \tag{14}$$
where C_D is the drag coefficient ($C_D \approx 0.02 \div 0.2$).

The air drag, D_c, of main cable and air drag, D_{tr}, of the transmission cable is
$$D_c = 0.5 C_{d,c} \rho V^2 H d_c, \quad D_{tr} = 0.5 C_{d,r} V^2 H d_{tr}, \quad [N], \tag{15}$$
where $C_{d,c}$ - drag coefficient of main cable, $C_{d,c} \approx 0.05 - 0.15$; H is rotor altitude, m; d_c is diameter of the main cable, m. $C_{d,r}$ - drag coefficient of the transmission cable, $C_{d,r} \approx 0.05 - 0.15$; d_{tr} is diameter of the transmission cable, m. Only half of this drag must be added to the total drag of wind installation:
$$D \approx D_r + D_w + D_d + 0.5 D_c + 0.5 D_{tr}, \quad [N] \tag{16}$$

If the wind installation is supported by dirigible, the lift force and air drag of dirigible must be added to wing lift force and total of system. The useful specific lift force of dirigible is about 5 N/m^3 (0,5 kg/m^3) at $H = 0$ and zero at $H = 6$ km. Full lift force is:
$$L = L_w + L_d - Mg - 0.5g(m_c + m_{tr}), \quad [N]. \tag{17}$$

Here M is total mass of installation (electron injectors + parachute/collector + half of cable and wires weight), kg; $g = 9.81$ m/s^2 is Earth acceleration. Lift force of dirigible $L_d \approx 5 U_d$ [N], where U_d is dirigible volume, m^3.

The mass of main and transmission cable are:
$$m_c = \gamma_c S_c L, \quad m_{tr} = 2\gamma_{tr} S_{tr} L, \quad [kg], \tag{18}$$
where γ_c is specific weight/density of cables, kg/m^3, $\gamma_c \approx 1500 \div 1800$ kg/m^3; S_c is cross section area of cables, m^2; L is length of cable, m.

The average angle α of connection line to horizon is
$$\sin \alpha \approx L/D, \tag{19}$$

The annual energy produced by the wind energy extraction installation equals
$$E = 8.33 N \quad [kWh]. \tag{20}$$

Project

Let us assume: on the wall of a seven story building is installed electron collector $A = 30 \times 60$ m $= 1800$ m^2. The wall of this building may be colored by conductive paint. The electron injectors are installed in front of wall (collector), in the distance of $L = 30$ m. Wind is perpendicular to the collector and has standard average permanent speed $V = 6$ m/s. The electron generator in optimal regime has the following data:
The power in efficiency $\eta = 0.25$:

$$N = 0.5\eta\rho AV^3 = 0.5 \cdot 0.25 \cdot 1.225 \cdot 1800 \cdot 6^3 \approx 100 \text{ kW}. \tag{21}$$

Optimal intensity of electric field:

$$E = 1.67 \cdot 10^3 V = 1.67 \cdot 10^3 \cdot 6 \approx 10 \text{ kV/m}. \tag{22}$$

Voltage and electric currency:

$$U = FL = 10 \cdot 30 = 300 \text{ kV}, \quad I = N/U = 100/300 = 0.333 \text{ A}. \tag{23}$$

Produced voltage is high, but a special electric capacitor converts the high voltage in low voltage.

Conclusion

Relatively no progress has been made in wind energy technology in the last years. While the energy from wind is free, its production is more expensive than its production in conventional electric power stations. Conventional wind energy devices have approached their maximum energy extraction potential relative to their installation cost. Current wind installations cannot significantly decrease a cost of kWh, provide the stability of energy production. They cannot continue significantly increase the power of single energy units.

The renewable energy industry needs revolutionary ideas that improve performance parameters (installation cost and power per unit) and that significantly decrease (by 5-10 times) the cost of energy production. The electron wind installations delineated in this paper can move the wind energy industry from stagnation to revolutionary potential.

The following is a list of benefits provided by the proposed new electron wind systems compared to current grown installations:

8. The produced energy is at least 0 times cheaper than energy produced in conventional electric stations which includes current wind installation.
9. The proposed system is relatively inexpensive (no expensive tower), it can be made with a very large collector thus capturing wind energy from an enormous area (tens of times more than typical wind turbines).
10. The proposed airborne electron installation does not require large ground space.
11. The installation may be located near customers.
12. Neither noise nor marring the landscape ruining the views.
13. The airborne energy production at high altitude is more stable because the wind is steadier. The wind may be zero near the surface but it is typically strong and steady at higher altitudes. This can be observed when it is calm on the ground, but clouds are moving in the sky. There are a strong permanent air streams at a high altitude at many regions of the USA and World.
14. The high altitude installation can be easy relocated to other places.
15. Offered installations are suitable for city having many high rise building.
16. The EABGs are invisible for population.
17. Offered installations produce high voltage direct electricity. That is advantage if energy is transferring in long distance.
18. Offered system is very suitable for airborne wind installation, because it is very light.
19. Offered system may be used as break and short power electric system of aircraft.

As with any new idea, the suggested concept is in need of research and development. The theoretical problems do not require fundamental breakthroughs. It is necessary to design small, cheap installations to study and get an experience in the design electron wind generator.

This paper has suggested some design solutions from patent application [2]. The author has many detailed analysis in addition to these presented projects. Organizations or investors are interested in these projects can address the author (http://Bolonkin.narod.ru , aBolonkin@juno.com , abolonkin@gmail.com).

The other ideas are in [1]-[6].

References

(Reader can find part of these articles in WEBs: http://Bolonkin.narod.ru/p65.htm, http://www.scribd.com(23); **Error! Hyperlink reference not valid.** , (45); http://www.archive.org/ (20) and http://aiaa.org (41) search "Bolonkin").

20. Bolonkin A.A., Utilization of Wind Energy at High Altitude, AIAA-2004-5756, AIAA-2004-5705. International Energy Conversion Engineering Conference at Providence, RI, USA, Aug.16-19, 2004.
21. Bolonkin, A.A., "Method of Utilization a Flow Energy and Power Installation for It", USA patent application 09/946,497 of 09/06/2001.
22. Bolonkin, A.A., Flight Wind Turbines. http://www.scribd.com/doc/138350864/
23. Bolonkin, A.A., "New Concepts, Ideas, Innovations in Aerospace, Technology and the Human Sciences", NOVA, 2006, 510 pgs. http://www.scribd.com/doc/24057071 , http://www.archive.org/details/NewConceptsIfeasAndInnovationsInAerospaceTechnologyAndHumanSciences;
24. Bolonkin, A.A., "New Technologies and Revolutionary Projects", Lulu, 2008, 324 pgs, http://www.scribd.com/doc/32744477 , http://www.archive.org/details/NewTechnologiesAndRevolutionaryProjects,
25. Bolonkin, A.A., Cathcart R.B., "Macro-Projects: Environments and Technologies", NOVA, 2007, 536 pgs. http://www.scribd.com/doc/24057930 . http://www.archive.org/details/Macro-projectsEnvironmentsAndTechnologies
26. Gipe P., Wind Power, Chelsea Green Publishing Co., Vermont, 1998.
27. Thresher R.W. and etc, Wind Technology Development: Large and Small Turbines, NRFL, 1999.
28. Galasso F.S., Advanced Fibers and Composite, Gordon and Branch Scientific Publisher, 1989.
29. Carbon and High Performance Fibers Directory and Data Book, London-New York: Chapmen& Hall, 1995, 6th ed., 385 p.
30. Concise Encyclopedia of Polymer Science and Engineering, Ed. J.I.Kroschwitz, N.Y.,Wiley,1990,1341p.
31. Dresselhaus, M.S., Carbon Nanotubes, by, Springer, 2000.
32. Joby turbines. http://www.jobyenergy.com/tech.
33. Makani turbine: http://theenergycollective.com/energynow/69484/airborne-wind-turbine-could-revolutionize-wind-power , http://www.treehugger.com/wind-technology/future-wind-power-9-cool-innovations.html .
34. Cost of renewable energy. http://www.irena.org/DocumentDownloads/Publications/RE_Technologies_Cost_Analysis-WIND_POWER.pdf
35. Koshkin P., Shirkevuch M., Directory of Elementary Physics., Moscow, Nauka, 1982 (in Russian).
36. Wikipedia. Wind Energy.

5 June 2013

Chapter 7

Electron Air Hypersonic Propulsion
Abstract.

Aviation, in general, and aerospace in particular needs new propulsion systems which allow the craft to reach high speeds by cheaper and more efficient methods. Author offers a new propulsion system using electrons for acceleration of the craft and having a high efficiency. As this system does not heat the air, it does not have the heating limitations of conventional air ramjet hypersonic engines. Offered engine can produce a thrust from a zero flight speed up to the desired space apparatus speed. It can work in any planet atmosphere (gas, liquid) and at very high altitude. The system can use apparatus surface for thrust and braking. For energy the system uses high voltage electricity which is not a problem if you have an appropriate electrostatic generator connected with any suitable engine.

Key words: Electron propulsion, EABP, hypersonic propulsion, space propulsion.

1. INTRODUCTION

Currently, turbo-rocket engines are widely used in aviation. Although they are good for subsonic speed, they are worse for small ($M < 2 \div 3$) supersonic speed and has tremendous difficulties achieving hypersonic speed ($4 < M < 6$). The current designs of ramjet hypersonic engines using high temperature compressed air are limited because current materials cannot withstand any greater temperature. Another significant limitation is that aircraft must use complex expensive hydrogen fuel [1]-[17].

A **jet engine** is a reaction engine that discharges a fast moving jet which generates thrust by *jet propulsion* in accordance with Newton's laws of motion. This broad definition of jet engines includes turbojets, turbofans, rockets, ramjets, and pulse jets. In general, most jet engines are internal combustion engines.

In common parlance, the term *jet engine* loosely refers to an internal combustion air breathing jet engine (a *duct engine*). These typically consist of an engine with a rotary (rotating) air compressor powered by a turbine ("Brayton cycle"), with the leftover power providing thrust via a propelling nozzle. These types of jet engines are primarily used by jet aircraft for long-distance travel. Early jet aircraft used turbojet engines which were relatively inefficient for subsonic flight. Modern subsonic jet aircraft usually use high-bypass turbofan engines which offer high speed with fuel efficiency comparable (over long distances) to piston and propeller aeroengines [18].

Electrostatic generators operate by using manual (or other) power to transform mechanical work into electric energy. Electrostatic generators develop electrostatic charges of opposite signs rendered to two conductors, using only electric forces, and work by using moving plates, drums, or belts to carry electric charge to a high potential electrode. The charge is generated by one of two methods: either the triboelectric effect (friction) or electrostatic induction.

2. INNOVATIONS

One simple version of the offered electronic ramjet propulsion engine (EABP) is shown in fig.1. Engine contains the tube. The ejectors of electrons 2 are installed in the entrance of the tube. The collector of electrons (grille) 3 is installed in the end of tube. The electric circle having the battery (electrostatic generator) 4 and regulator of voltage 7 connects the ejector and grille.

The engine works the following way. The ejectors eject the electrons into tube. The strong electric field between injectors and grill moves them to grill. Electrons push (accelerate) the air to tube exit. When the electrons reach the grill, they enter the grill and close the electric circuit. The accelerated air (air jet) with high speed flows out from engine and creates the thrust. In correct design engine this thrust may be enough for moving the craft.

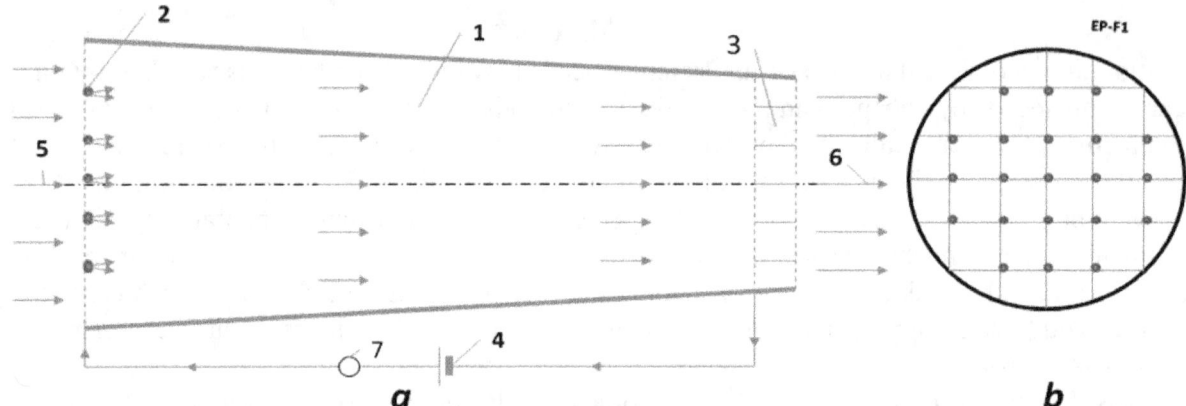

Fig.1. Electron ramjet engine (EABP). *a* – side view, *b* – forward view. *Notations:* 1 – engine; 2 – injector of electrons; 3 – collector of electrons; 4 – electric issue; 5 – enter air; 6 – exit air jet; 7 – regulator of an electric voltage (electron regulator).

The proposed idea of a propulsion engine has many versions. One of them is shown in fig. 2a. That is a conventional fuselage or wing (in fig. 2a it is shown the gross section of the wing). The electron injectors are installed in beginning of the fuselage (wing) surface. The collectors are installed in the end of the fuselage/wing. The electrons accelerate the air around the flying apparatus and the electric forces produce the thrust.

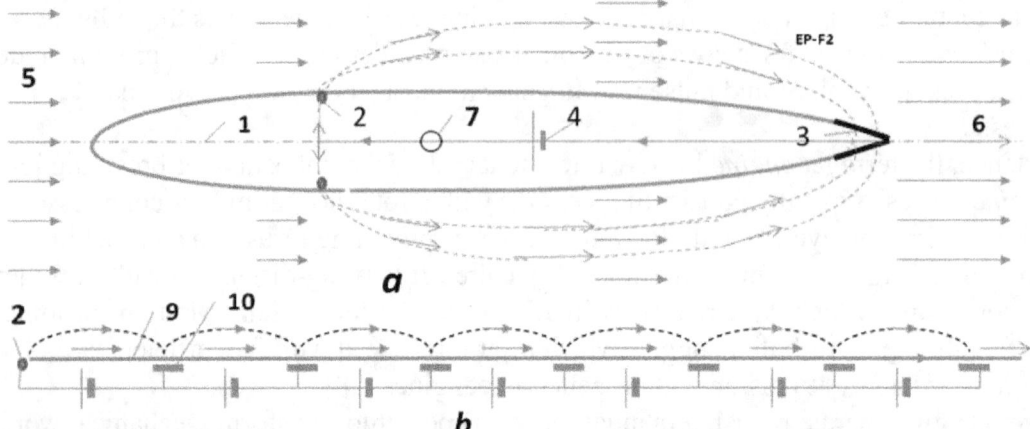

Fig.2. Outer Electron ramjet engine (EABP). *a* – side view of the fuselage or a gross-section of wing, *b* – surface electron engine. *Notations:* 1 – fuselage or wing; 2 – injector of electrons; 3 – collector of electrons; 4 – electric issue; 5 – enter air; 6 – exit air jet; 7 – electric (electron) regulator; 9 –surface (isolator) of fly apparatus; 10 – electric plate.

One possible electric schema of the proposed engine, shown in fig. 3, has an additional closed loop electric circles which allows extracting the electrons from main electric circle and collecting electrons from air flow to back into main circle, to heat the electron ejectors (cathodes) if it is necessary.

Principal differences the offered EABP engines from known propulsion systems/engines.

From air-breathing engine:
1. Air-breathing propulsion engine as any heat engine compresses and HEATS the air.
   The electronic propulsion engine does not compress and does not heat the entered air.

2. Air-breathing propulsion engine expends liquid fuel.
   The electronic propulsion engine expends electric energy.

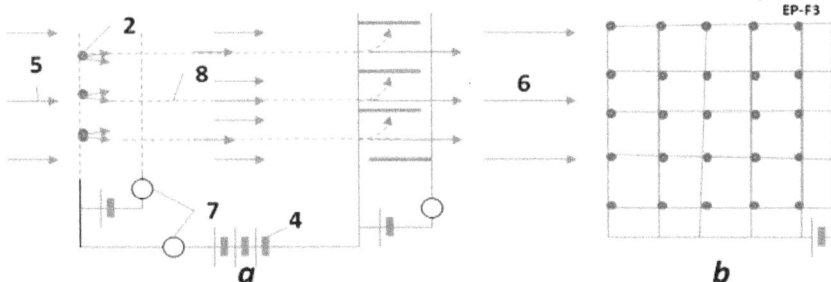

Fig.3. The electrical circuit of one version EABP engine. Notations are same with figs. 1 – 2. *a* is side view, *b* is forward view.

From rockets:
1. Rocket expends fuel.
   The electronic propulsion engine expends electric energy.
   From the electric rocket engine.
1. The electric rocket engines and the electronic propulsion work in different mediums. The electronic propulsion uses the outer medium (atmosphere, gas, liquid, etc.) while most electric rockets may work only in vacuum.
2. The electric rocket engines can use only positive ions.
   The electronic propulsion system use only electrons.
3. The electric rocket engines expends the apparatus mass (for example, plasma).
   The electronic propulsion system does NOT expends the apparatus mass.

Advantages and disadvantages of the proposed electron propulsion system in comparison with the conventional air propulsion systems.

The suggested new propulsion principle has the following advantages and disadvantages in comparison with conventional air-breathing engine propulsion systems used at present time.
   Advantages:
1. All current air-breathing propulsion engines as any heat engine compresses and HEATS the air.
   As the result the heat efficiency is about 30% or low.
   The electronic propulsion engine (EABP) does not compress and does not heat the entered air.
   His electric efficiency is about 100% which makes it 3 more times efficient.
2. All current the air-breathing engines has small efficiency in hypersonic speed ($3 < M < 5$), because the
   high compressed air has big temperature and current material cannot keep them. Conventional
   hypersonic engine is very complex, needs hydrogen fuel. There is no production of the hypersonic engine at present time although its research and design is doing about 20 years. For $M > 6$ the heat hypersonic engine cannot work.
   The electron engine not heat an air and can work at any speed. That means one may be used as a cheap space launcher and engine of the super speed aircraft.
3. The electronic engine is very simple and cheap.
4. The outer air ship surface may be used as engine. The aircraft may not have nacelles (moto-gondols).
   That means high aerodynamic efficiency of flight apparatus.
5. The outer surface electronic engine (fig.2b) may be used for creating the laminar boundary layer.
   That means low (minimal) air friction and very high aerodynamic efficiency of flight apparatus.
6. The outer surface electronic engine (fig.2b) may be used for creating the high lift force.

That means a low landing speed, decreasing the take-off and landing distances, VTOL aircraft.
7. The electron engines can work at very high atmosphere.
8. The EABP can works in any atmosphere and in other planets; space apparatus can use any matter of planets, asteroids and apparatus garbage in the EABP engine.

Possible Disadvantages:

1. Main disadvantage of electron propulsion engine: the aircraft needs strong high voltage electric power. This problem may be solved by connecting the conventional engine with static electric generator. The static electric generator is lightweight and cheap. Electrostatic generator must be researched and developed in order for it to produce high voltage direct electricity. One, although not suitable for use by population and industry, but the electrostatic generators are needed for electron propulsion engine needed in very high voltage (up 2 millions volts).

3. THEORY OF ELECTRON PROPULSION (EABP). COMPUTATION AND ESTIMATION.

1. Thrust of EABP. The thrust of the jet electron engine is (we use the Law of Impulse):

$$T = m(V_f - V) = m\Delta V, \quad m = \rho SV, \quad T = \rho SV\Delta V, \quad T_s = \rho V\Delta V, \quad (1)$$

where T is thrust, N; m is air mass passed through engine in one second, kg/s; V_f is an exit speed of air (medium), m/s; V is an entry speed of air (medium), (flight speed of the apparatus), m/s; ΔV is increasing of air (medium) speed into engine, m/s; ρ is air (medium) density, kg/m³; S is ender area of engine, m²; T_s is specific thrust of engine, N/m².

The energy A_t [J] getting by flight apparatus from thrust is

$$A_t = TVt, \quad (2)$$

where t is time, sec.

From other hand, the energy A_e [J] getting from of electric current is

$$A_e = UIt, \quad (3)$$

where U is voltage between entrance and exit of engine, V; I is electric current, A.

The heat efficiency of the EABP is close to 1, because no heating of air into engine (the increasing the speed of all air mass is in one direction by electric field).
That way

$$A_t \approx A_e. \quad (4)$$

From (1) – (4) and $I_s = I/S$ we get ($V \neq 0$)

$$T_s = \frac{U}{V} I_s, \quad \Delta V = \frac{UI_s}{\rho V^2}, \quad (5)$$

where I_s is density of electric currency about apparatus, A/m², ΔV is increasing air (medium) speed into engine, m/s.

Example 1. Let us take the $U = 10^6$ V, $I_s = 10$ A/m², flight speed $V = 200$ m/s, $\rho = 1$ kg/m³. Then $T_s = 5 \times 10^4$ N/m² = 5 tons/m², $\Delta V = 250$ m/s.

Example 2. Let us take the $U = 4 \times 10^6$ V, $I_s = 100$ A/m², flight speed $V = 8000$ m/s, $\rho = 1$ kg/m³. Then $T_s = 5 \times 10^4$ N/m² = 5 tons/m², $\Delta V = 6.25$ m/s.

The same way we can get the request power and getting thrust when the flight speed equals zero:

$$P_s = 0.5m\Delta V^2, \quad m = \rho\Delta V, \quad T_s = P_s/\Delta V, \quad P_s = 0.5\rho\Delta V^3, \quad T_s = 0.5\rho\Delta V^2, \quad (6)$$

where P_s is electric power for 1 m², W/m²; ΔV is increasing air speed into engine, m/s; m is air exemption mass passed throw engine in one second, kg/s;

Example 3. Let us take the $U = 10^6$ V, $I_s = 10$ A/m², exit speed $\Delta V = 100$ m/s, $\rho = 1$ kg/m³. Then the start thrust is $T_s = 10^5$ N/m² = 10 tons/m² if the start power is $P_s = 10^7$ W/m².

2. Efficiency of Electron EABP engine.

Efficiency η of any jet (air flight) propulsion is production of two values: propulsion efficiency η_p and engine (cycle) efficiency η_e:

$$\eta = \eta_p \eta_e, \quad \text{where} \quad \eta_p = V/(V + 0.5 \Delta V). \tag{7}$$

The flight efficiency for heat and electronic propulsion are same. They depend only on ΔV. But thermodynamic (cycle) efficiency of the heat engine is low about 25 ÷ 35%. The heat engine looses a great deal of energy from the hot exit jet. For high speed over $M > 3$ the conventional air rocket (jet) engine looses efficiency very quickly. The aviation designers try to use the hydrogen fuel, but after $M > 5$ the hydrogen fuel is also useless. The offered electronic jet engine accelerates air by electricity. It has efficiency close to 100% as the only loss of energy is the extraction of the electrons from cathode and ionizations of air molecules. This energy is about tens electron-volts (eV). The energy spent for acceleration of the air molecules by electrons/ions is hundreds of thousands of eV. That means the total efficiency of EABP is 3 times more than conventional air jet propulsion.

The second very important point: efficiency of EABP does not depend upon speed of apparatus.

The other advantages: we can make a very large entrance area of engine, we can use the fuselage and wings, stabilizer and keel of plane as engine.

3. Electron speed.
The electron speed about the gas (air) jet may be computed by equation:

$$j_s = qn_{-}b_{-}E + qD_{-}(dn_{-}/dx), \tag{8}$$

where j_s is density of electric currency about jet, A/m²; $q = 1.6 \times 10^{-19}$ C is charge of single electron, C; n_{-} is density of injected electrons (negative charges) in 1 m³; b_{-} is charge mobility of negative charges, m²/sV; E is electric intensity, V/m; D_{-} is diffusion coefficient of charges; dn_{-}/dx is gradient of charges. For our estimation we put $dn_{-}/dx = 0$. In this case

$$j_s = qn_{-}b_{-}E, \quad Q = qn_{-}, \quad v = bE, \quad j_s = Qv, \tag{9}$$

where Q is density of the negative charge in 1 m³; v is speed of the negative charges about jet, m/s.

The negative charge mobility for normal pressure and temperature $T = 20°C$ is:

$$\text{In dry air } b_{-} = 1.9 \times 10^{-4} \text{ m}^2/\text{sV}, \text{ in humid air } b_{-} = 2.1 \times 10^{-4} \text{ m}^2/\text{sV}. \tag{10}$$

In diapason of pressure from 13 to 6×10^6 Pa the mobility follows the Law bp = const, where p is air pressure. When air density decreases, the charge mobility increases. The mobility stregnth depends upon the purity of gas.

For normal air density the electric intensity must be less than 3 MV ($E < 3$ MV). Otherwise the electric breakdown may be:

If $v > 0$, the electrons accelerate the air into engine ($E > 0$ and engine spend energy). If $v < 0$, the electrons beak the air into engine ($E < 0$ and engine can produce energy). If $v = 0$ (electron speed about apparatus equals V), the electric resistance of jet into engine is zero.

Example 4. If $E = 10^6$ than $v = 200$ m/s.

4. Electron injectors.

There are some methods for getting the electron emissions: hot cathode emission, cold field electron emission (edge cold emission, edge cathode), photo emission, radiation emission, radioisotope emission and so on. We consider only the hot emission and shortly the cold field electron emission (edge cathodes).

The **hot cathode** emission computed by equation:

$$j_s = BT^2 exp(-A/kT), \quad (11)$$

where B is coefficient, A/cm^{2}K^2; T is catode temperature, K; $k = 1.38 \times 10^{-23}$ [J/K] is Bolzmann constant; A is thermoelectron exit work, eV. Both values A, B depend from material of cathode and its cover. The "A" changes from 1.6 to 5 eV, the "B" changes from 0.5 to 120 A/cm^{2}K^2. Boron thermo-cathode produces electric currency up 200 A/cm^2. For temperature $1400 \div 1500$K the cathode can produce currency up 1000 A/cm^2. The life of cathode can reach some years [19]-[20].

The edge cold emission. The cold field electron emission uses the edge cathodes. It is known that the electric intensity E_e in the edge is

$$E_e = U/a . \quad (12)$$

Here a is radious of the edge. If voltage between the edge and nears net (anode) is $U = 1000$ V, the radius of edge $a = 10^{-5}$ m, electric intensity at edge is the $E_a = 10^8$ V/m. That is enough for the electron emission. The density of electric current may reach up 10^4 A/cm^2. For getting the required currency we make the need number of edges.

4. SUMMARY AND DISCUSSION.

The author proposed the principally new propulsion system (engine) using the outer medium (air) and electric energy. It is not comparable to conventional heat propulsion because the heat jet engine gets the thrust by compressing the air, burning the fuel into air, heating, accelerating the hot air and expiring the hot gas in atmosphere.

The offered EABP engine is accelerating the air (medium) by a principally new method – by electric field which does not need atmospheric oxygen and thus can work in any atmosphere of other planets. This engine does not require compressing and heating of medium and, as such, does not have limitations of high temperature, high flight speed and rare atmosphere.

This engine is also dissimilar to known space electric engines. The space electric engine takes an extracted mass from itself, ionizes it, and accelerates springing forward in a vacuum. It has very small thrust, works poorly into any atmosphere and works worse if the atmosphere has a high density. The EABP does not take the extracted mass, can work only in atmosphere and works better if the atmosphere has a high density.

The main disadvantage of the offered engine is the requirement of high voltage electricity. For getting the electricity may be used the conventional internal turbo engine connected with electro-statics generator. Electro-statics power generator is light-weight and produces high voltage electricity.

The researches having relation to this topic are presented in [1]-[17].

References

[1]. A.A. Bolonkin, "High Speed Catapult Aviation", AIAA-2005-6221, presented to *Atmospheric Flight Mechanic Conference* – 2005. 15–18 August, USA.

[2]. A.A. Bolonkin, "Air Cable Transport System", *Journal of Aircraft*, Vol. 40, No. 2, July-August 2003, pp. 265–269.

[3]. A.A. Bolonkin, "Bolonkin's Method Movement of Vehicles and Installation for It", US Patent

6,494,143 B1, Priority is on 28 June 2001.

[4]. A.A. Bolonkin, "Air Cable Transport and Bridges", TN 7567, *International Air & Space Symposium – The Next 100 Years*, 14-17 July 2002, Dayton, Ohio, USA

[5]. A.A. Bolonkin, "Non-Rocket Missile Rope Launcher", IAC-02-IAA.S.P.14, *53rd International Astronautical Congress, The World Space Congress – 2002*, 10–19 Oct 2002, Houston, Texas, USA.

[6]. A.A. Bolonkin, "Inexpensive Cable Space Launcher of High Capability", IAC-02-V.P.07, *53rd International Astronautical Congress. The World Space Congress – 2002*, 10–19 Oct. 2002. Houston, Texas, USA.

[7]. A.A. Bolonkin, "Non-Rocket Space Rope Launcher for People", IAC-02-V.P.06, *53rd International Astronautical Congress. The World Space Congress – 2002*, 10–19 Oct 2002, Houston, Texas, USA.

[8]. A.A. Bolonkin, "*Non-Rocket Space Launch and Flight*", Elsevier, 2005, 468 pgs. Attachment 2: High speed catapult aviation, pp.359-369. http://www.scribd.com/doc/24056182, http://www.archive.org/details/Non-rocketSpaceLaunchAndFlight ,

[9]. A.A. Bolonkin, "*New Concepts, Ideas, Innovations in Aerospace, Technology and the Human Sciences*", NOVA, 2006, 510 pgs. http://www.scribd.com/doc/24057071 , http://www.archive.org/details/NewConceptsIfeasAndInnovationsInAerospaceTechnologyAndHumanSciences

[10]. A.A. Bolonkin, R. Cathcart, "*Macro-Projects: Environments and Technologies*", NOVA, 2007, 536 pgs. http://www.scribd.com/doc/24057930 . http://www.archive.org/details/Macro-projectsEnvironmentsAndTechnologies .

[11]. A.A. Bolonkin, *Femtotechnologies and Revolutionary Projects*. Lambert, USA, 2011. 538 p., 16 Mb. http://www.scribd.com/doc/75519828/ , http://www.archive.org/details/FemtotechnologiesAndRevolutionaryProjects

[12]. A.A. Bolonkin, *LIFE. SCIENCE. FUTURE* (Biography notes, researches and innovations). Scribd, 2010, 208 pgs. 16 Mb. http://www.scribd.com/doc/48229884, http://www.archive.org/details/Life.Science.Future.biographyNotesResearchesAndInnovations

[13]. A.A. Bolonkin, *Universe, Human Immortality and Future Human Evaluation*. Scribd. 2010г., 4.8 Mb. http://www.archive.org/details/UniverseHumanImmortalityAndFutureHumanEvaluation, http://www.scribd.com/doc/52969933/

[14]. A.A.Bolonkin, "Magnetic Space Launcher" has been published online 15 December 2010, in the ASCE, *Journal of Aerospace Engineering* (Vol.24, No.1, 2011, pp.124-134). http://www.scribd.com/doc/24051286/

[15]. A.A.Bolonkin, Universe. Relations between Time, Matter, Volume, Distance, and Energy (part 1) http://viXra.org/abs/1207.0075, http://www.scribd.com/doc/100541327/ , http://archive.org/details/Universe.RelationsBetweenTimeMatterVolumeDistanceAndEnergy

[16]. A.A.Bolonkin, Lower Current and Plasma Magnetic Railguns. Internet, 2008. http://www.scribd.com/doc/31090728 ; http://Bolonkin.narod.ru/p65.htm .

[17] A.A.Bolonkin, Electrostatic Climber for Space Elevator and Launcher. Paper AIAA-2007-5838 for *43 Joint Propulsion Conference*. Chincinnati, Ohio, USA, 9 – 11 July,2007. See also [10], Ch.4, pp. 65-82.

[18] W.J. Hesse and el. Jet Propulsion for Aerospace Application, Second Edition, Pitman Publishing Corp. NY.

[19] N.I. Koshkin and M.G. Shirkebich, Directory of Elementary Physics, Nauka, Moscow, 1982 (in Russian).

[20] I.K. Kikoin. Table of Physics values. Atomisdat, Moscow, 1976 (in Russian).

May 27, 2013

Chapter 8
Electric Hypersonic Space Aircraft
Alexander Bolonkin
C&R, USA abolonkin@juno.com

Abstract

Aviation, in general, and aerospace in particular needs new propulsion systems which allow a craft to reach high speeds by cheaper and more efficient methods. Author offers a new high efficiency propulsion system using electrons for acceleration of the craft. As this system does not heat the air, it does not have the heating limitations of conventional air ramjet hypersonic engines. Offered engine can produce a thrust from a zero flight speed up to the desired escape velocity for space launch. It can work in any planet atmosphere (gas, liquid) and at high altitude. The system can use apparatus surface for thrust and braking. For energy the system uses high voltage electricity which is not a problem if you have an appropriate electrostatic generator connected with any suitable engine. The new propulsion system applies to hypersonic long-range aviation, for launch of space craft and as a high efficiency rocket in solar space. This can be actualized using current technology.

Key words: Electron propulsion, hypersonic propulsion, space propulsion, ABEP.

1. INTRODUCTION

Let us consider the status of the problem succinctly.

Aviation and space launch. In the last half century, development of aviation and space launch proceeded very slowly. The last major advance in aviation was the introduction of the reactive engine. Space launch started using a chemical rocket and is still using it at the present time. For more than thirty years, employing the same old engines researchers unsurprisingly come up against the same barriers which do not allow significantly improving flight. Specifically, supersonic aircraft has high fuel consumption and chemical rocket engine is limited by the chemical energy of the rocket fuel.

Currently, turbojet engines are widely used in aviation. Although they are good for subsonic speed, they are worse for small ($M < 2 \div 3$) supersonic speed and has tremendous difficulties achieving hypersonic speed ($4 < M < 6$). The current designs of ramjet hypersonic engines using high temperature compressed air are limited because current materials cannot withstand any greater temperature. Another significant limitation is that hypersonic aircraft must use complex expensive hydrogen fuel [1]-[19].

A jet engine is a reaction engine that discharges a fast moving jet which generates thrust by *jet propulsion* in accordance with Newton's laws of motion. This broad definition of jet engines includes turbojets, turbofans, rockets, ramjets, and pulse jets. In general, most jet engines are internal combustion engines.

In common parlance, the term *jet engine* loosely refers to an internal combustion air breathing jet engine (a *duct engine*). These typically consist of an engine with a rotary (rotating) air compressor powered by a turbine ("Brayton cycle"), with the leftover power providing thrust via a propelling nozzle. These types of jet engines are primarily used by jet aircraft for long-distance travel. Early jet aircraft used turbojet engines which were relatively inefficient for subsonic flight. Modern subsonic jet aircraft usually use high-bypass turbofan engines which offer high speed with fuel efficiency comparable (over long distances) to piston and propeller aero-engines [24].

Hypersonic transport. While conventional turbo and ramjet engines are able to remain reasonably efficient up to Mach 5.5, some ideas for very high-speed flight above Mach 6 are also sometimes discussed, with the aim of reducing travel times down to one or two hours anywhere in the world.

These vehicle proposals very typically either use rocket or scramjet engines; pulse detonation engines have also been proposed. There are many difficulties with such flight, both technical and economic.

Rocket-engine vehicles, while technically practical (either as ballistic transports or as semiballistic transports using wings), use a very large amount of propellant and operate best at speeds between about Mach 8 and orbital speeds. Rockets compete best with air-breathing jet engines on cost at very long range; however, even for antipodal travel, costs would be only somewhat lower than orbital launch costs.

Scramjets currently are not practical for passenger-carrying vehicles due to technological limitations.

Ion wind, **ionic wind**, **coronal wind** or **electric wind** are expressions formerly used to describe the resulting localized neutral flow induced by electrostatic forces linked to Corona discharge arising at the tips of some sharp conductors (such as points or blades) submitted to high-voltages relative to ground. Modern implementations belong to the family of Electrohydrodynamic (EHD) devices. Ion wind production machines can be now considered as Electrohydrodynamic (EHD) pumps. Francis Hauksbee, curator of instruments for the Royal Society of London, made the earliest report of electric wind in 1709.

An **ionocraft** or **ion-propelled aircraft** (commonly known as a **lifter** or **hexalifter**) is a device that uses an electrical electrohydrodynamic (EHD) phenomenon to produce thrust in the air without requiring any combustion or moving parts. The term "Ionocraft" dates back to the 1960s, an era in which EHD experiments were at their peak. In its basic form, it simply consists of two parallel conductive electrodes; one in the form of a fine wire and another which may be formed of either a wire grid, tubes or foil skirts with a smooth round surface. When such an arrangement is powered up by high voltage in the range of a few kilovolts, it produces small thrust. The ionocraft forms part of the EHD thruster family, but is a special case in which the ionisation and accelerating stages are combined into a single stage. The device is a popular science fair project for students. It is also popular among anti-gravity or so-called "electrogravitics" proponents, especially on the Internet. The term "lifter" is an accurate description because it is not an anti-gravity device, but produces lift in the same sense as a rocket from the reaction force from driving the ionized air downward. Much like a rocket or a jet engine (it can actually be much more thrust efficient than a jet engine). The force that an ionocraft generates is oriented consistently along its own axis regardless of the surrounding gravitational field. Claims of the device working in a vacuum also have been disproved.

In its basic form, the ionocraft is able to produce forces great enough to lift about a gram of payload per watt, so its use is restricted to a tethered model. Ionocraft capable of payloads in the order of a few grams usually need to be powered by power sources and high voltage converters weighing a few kilograms, so although its simplistic design makes it an excellent way to experiment with this technology, it is unlikely that a fully autonomous ionocraft will be made with the present construction methods. This area has not been researched with good ideas, theory, design and experiment of ionocraft.

This article offers the new theory and principal design of the new engine, propulsion system for aviation, space launch and flight. These ideas include the new lightweight electrostatic high voltage electric generators. At present time **Electrostatic generators** operate by using manual (or other) power to transform mechanical work into electric energy. Electrostatic generators develop electrostatic charges of opposite signs rendered to two conductors, using only electric forces, and work by using moving plates, drums, or belts to carry electric charge to a high potential electrode. The charge is generated by one of two methods: either the triboelectric effect (friction) or electrostatic induction.

2. INNOVATIONS

One simple version of the offered electronic ramjet propulsion engine (ABEP) is shown in fig.1.

Engine contains the tube 2. The injectors of electrons 3 (or ions) are installed in the entrance of the tube. The second electrode-collector of electrons (ring, plats or net) 4 is installed in the end of tube. The electric circle having the battery (electrostatic generator) 8 and regulator of voltage connects the injectors 3 and back electrode (net, plats) 4. There is compensator 5 connected with forward 3 and back injectors 6 which discharges an excess charges in an exit flow. The charge compensator 5 is one of the most important innovations. All early proposed models of ion lifters cannot work without connection to Earth because they get self-charge and loss efficiency.

The engine works the following way. The injectors 3 eject the electrons (or ions) into tube (engine) 2. The strong electric field between injectors 3 and back electrode (ring, plats, net) 4 moves them to back electrode 4. Electrons (or ions) push (accelerate) the air to the tube exit. When the electrons (ions) reach the collector (electrode) 4, they (or part of them) enter the electrode and close the electric circuit. The excess part of charges is compensated by compensator 5. The accelerated air (air jet) with high speed flows out from engine and creates the thrust. In correctly designed engine this thrust may be enough for vertical start or moving the craft up to high hypersonic speed.

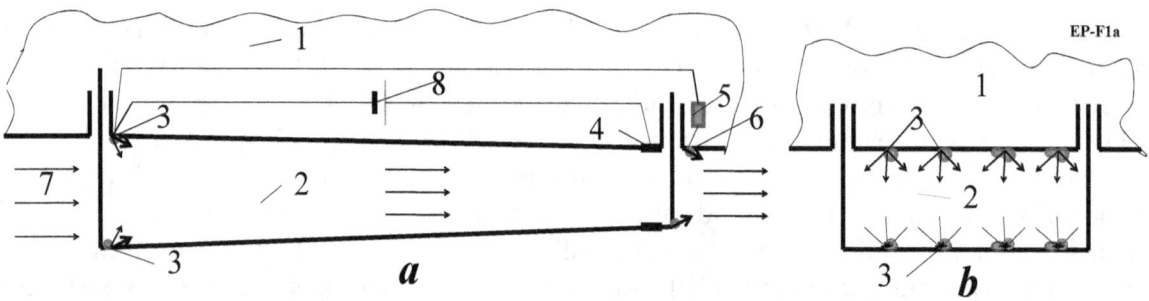

Fig.1. Electron ramjet engine (ABEP). *a* – side view, *b* – forward view. *Notations:* 1 – aircraft body, 2 – propulsion body, 3 – injector of charges (forward electrode), 4 – back electrode, 5 - separator (compensator) of charges; 6 – back injector of charges (opposed the forward injector 3), 7 – air flow. 8 – issue of high voltage (example, the electrostatic generator).

The proposed idea of a propulsion engine has many versions. One of them suitable for VTOL (aircraft with vertical start and lending) or helicopter is shown in fig. 2.

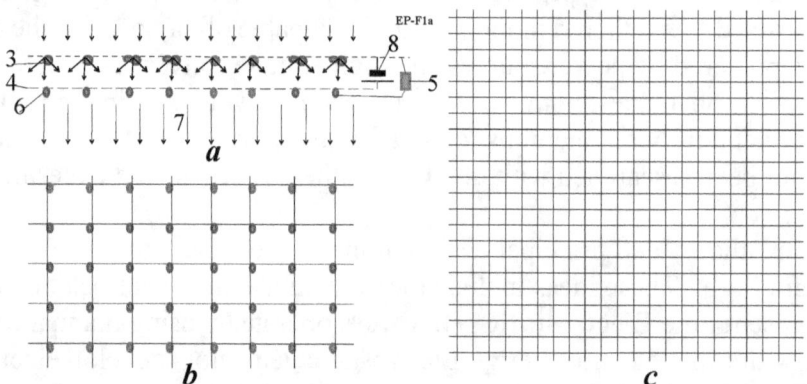

Fig.2. Electron ramjet engine (ABEP) for vertical start. *a* – side view, *b* –net of injector (top view), *c* – lover electrode (lower view). *Notations:* 3 – injector of charges (forward electrode), 4 – back electrode, 5 separator of charges; 6 – lower injector of charges (opposed the forward injector 3), 7 – air flow. 8 – source of high voltage (example, the electrostatic generator).

For economical vertical start and helicopter flight we need in the engine a large area for entrance and exit. This version has two nets: rare upper net with injectors (fig. 2b) and a denser mesh from a thin wire (fig.2c). These nets can be foldable and installed in the fuselage and under wings. The aircraft in disk form is a suitable form in this case (subsonic high speed aircraft – helicopter).

One possible electric schema of the proposed engine, shown in fig. 3, has an additional closed loop electric circles which allows more efficiency extracting the electrons from main electric circle and collecting electrons from air flow to back into main circle, to heat the electron injectors (cathodes) if it is necessary.

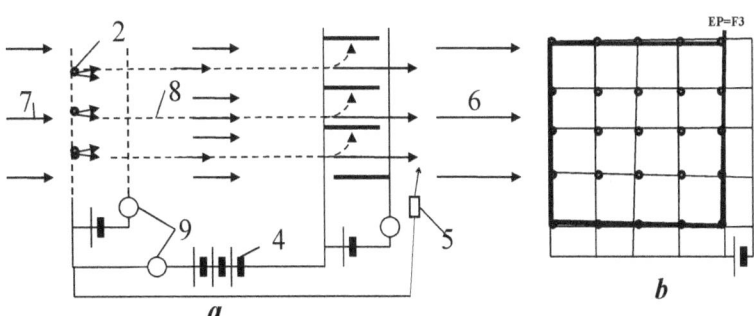

Fig.3. The electrical circuit of one version ABEP engine. Notations: *a* is side view, *b* is forward view; 2 – injector; 4 – electric issue, for example, the electrostatic generator, 5 – Compensator; 6 - exit flow, 7 – air flow, 8 – trajectory of the charges, 9 – regulator.

Principal differences the offered ABEP engines from known propulsion systems/engines.
From air-breathing engine:
1. Air-breathing propulsion engine as any heat engine compresses and HEATS the air. The electronic propulsion engine does not compress and does not heat the entered air.
2. Air-breathing propulsion engine expends liquid fuel. The electronic propulsion engine expends electric energy. But one may have the turbojet engine and electrostatic generator.

From rockets:
1. Rocket expends fuel.
   The electronic propulsion engine expends electric energy.

From the electric rocket engine.
1. The electric rocket engines and the electronic propulsion work in different mediums. The electronic propulsion uses the outer medium (atmosphere, gas, liquid, etc.) while most electric rockets may work only in vacuum. The ABEP has in tens-hundreds times more the ratio thrust/power. But ABEP can work in vacuum if one has the electric source.
2. The electric rocket engines can accelerate only positive ions.
   The electronic propulsion system accelerate electrons and positive ions.
3. The electric rocket engines expend the apparatus mass (special fuel).
   The electronic propulsion systems do NOT expend the apparatus mass or expends very few it. ABEP can accelerates (uses as passive fuel) ANY mass of space body (meteorites, asteroids, planets, planet satellites, dust, etc.)

From electrostatic lifter.
1. **The ABEP has correct design. In particular the ABEP has body, tube, new injector and the charge compensator.**
   As a result:
   1. The electrostatic lifter, using maximum voltage can produce only some grams of lift force and additional wind speed up 10 m/s. The ABEP can produce tons of thrust, accelerate air up hundreds m/s and works in flow having hypersonic speed. It may be used in both atmospheric as well as in space ships.
   2. The electrostatic lifter has very low efficiency. This is why the lifter had no application in aerospace although his idea is known for some hundreds of years and tested numerous times.

The ABEP has significantly higher efficiency (ratio thrust/power) than any current lifter because of our unique design.
3. ABEP works in hypersonic speed in atmosphere and can work in vacuum (outer space) using as passive propulsion any matter of the space bodies.

Advantages and disadvantages of the proposed electron propulsion system in comparison with the conventional air propulsion systems.

The suggested new propulsion principle has the following advantages and disadvantages in comparison with conventional air-breathing engine propulsion systems used at present time.

Advantages:
1. All current air-breathing propulsion engines as any heat engine compresses and HEATS the air. As a result the heat efficiency is about 30% or low. The electronic propulsion engine (ABEP) does not compress and does not heat the entered air. Its electric efficiency is about 100% which makes it 3 more times efficient.
2. All current the air-breathing engines has small efficiency in hypersonic speed ($3 < M < 5$), because the high compressed air has high temperature and current material cannot sustain them. Conventional hypersonic engine (for $M > 5$) is very complex, needs hydrogen fuel. There is no production of the hypersonic engine at present time although its research and design has been completed about 20 years. For $M > 6$ the heat hypersonic engine cannot work because the hot air (fuel) begins to dissociate and ionize. The electron engine does not heat the air and can work at any speed. That means it may be used as a cheap space launcher and engine of a super speed aircraft.
3. The electronic engine is very simple and cheap.
4. The outer air ship surface may be used as engine [1]. The aircraft may not have nacelles (moto-gondolas). That means higher aerodynamic efficiency of flight apparatus.
5. The outer surface electronic engine ([1] fig.2b) may be used for creating the laminar boundary layer. That means low (minimal) air friction and very high aerodynamic efficiency of flight apparatus.
6. The offered engine and outer surface electronic engine ([1] fig.2b) may be used for creating the high lift force. That means a lower landing speed, decreasing the take-off and landing distances, VTOL aircraft.
7. The electron engines can work at very high atmosphere and vacuum. If ABEP has an independent electric source (for example nuclear reactor), it operates with high efficiency (high exit speed) and can use as fuel (reaction mass) any matter of space bodies: space dust, meteorites, asteroids and planets, converted in the dust, liquid or gas.
8. The ABEP having source of electricity can works in any atmosphere and in other planets; space craft can use any matter of planets, asteroids and apparatus garbage in the ABEP engine.

Possible Disadvantages:
1. Main disadvantage of electron propulsion engine: the aircraft needs strong high voltage electric power. This problem may be solved by connecting the conventional engine with static electric generator. The static electric generator is lightweight and cheap. Electrostatic generator must be researched and developed in order for it to produce high voltage direct electricity. One, although not suitable for use by population and industry at present time, but the electrostatic generators are needed for electron propulsion engine needed in very high voltage (up 2 millions volts). High voltage electricity is more suitable for efficiency long distance transmission. The author has works which allow easy transfer the high voltage to low voltage of direct or variable electric currency (or back).

3. THEORY OF ELECTRON PROPULSION (ABEP). COMPUTATION AND ESTIMATION.

1. Thrust of ABEP. The thrust of the jet electron engine is (we use the Law of Impulse):

$$T = m(V_f - V) = m\Delta V, \quad m = \rho S V, \quad T = \rho S V \Delta V,$$
$$T_s = \rho V \Delta V, \quad T \approx Id/b \approx P/bE, \quad T_s \approx P_s/bE, \quad (1)$$

where T is thrust, N; m is air mass passed through engine in one second, kg/s; V_f is an exit speed of air (medium), m/s; V is an entry speed of air (medium), (flight speed of the apparatus), m/s; ΔV is increasing of air (medium) speed into engine, m/s; ρ is air (medium) density, kg/m³; S is entrance area of engine, m²; T_s is specific thrust of engine, N/m²; I is electric currency, A; d is distance between cathode and anode, m; b is mobility of charges, m²/sV (ions in air in atmospheric pressure has $b \approx 2 \cdot 10^{-4}$ m²/sV, where V is voltage in V); P is power of electricity, W; $E = U/d$ is intensity of an electric field, V/m; U is voltage between cathode and anode, V.

The energy A_t [J] getting by flight apparatus from thrust is

$$A_t = TVt, \quad (2)$$

where t is time, sec.

On other hand, the energy A_e [J] getting from of electric current is

$$A_e = UIt, \quad (3)$$

where U is voltage between entrance and exit of engine, V; I is electric current, A.

The heat efficiency of the ABEP is close to 1, because no heating of air into engine (the increasing the speed of all air mass is in one direction by electric field).

That way

$$A_t \approx A_e. \quad (4)$$

From (1) – (4) and $I_s = I/S$ we get ($V \neq 0$)

$$T_s = \frac{U}{V} I_s, \quad \Delta V = \frac{U I_s}{\rho V^2}, \quad (5)$$

where I_s is density of electric currency about apparatus, A/m²; ΔV is increasing air (medium) speed into engine, m/s.

Example 1. Let us take the $U = 10^6$ V, $I_s = 10$ A/m², flight speed $V = 200$ m/s, $\rho = 1$ kg/m³. Then $T_s = 5 \times 10^4$ N/m² = 5 tons/m², $\Delta V = 250$ m/s.

Example 2. Let us take the $U = 4 \times 10^6$ V, $I_s = 100$ A/m², flight speed $V = 8000$ m/s, $\rho = 1$ kg/m³. Then $T_s = 5 \times 10^4$ N/m² = 5 tons/m², $\Delta V = 6.25$ m/s.

The same way we can get the request power and getting thrust when the flight speed equals zero:
$$A_s = 0.5 m_s \Delta V^2, \quad m_s = \rho \Delta V, \quad T_s = m_s \cdot \Delta V = \rho \Delta V^2, \quad P_s = T_s \cdot 0.5 \Delta V = 0.5 \rho \Delta V^3, \quad (6)$$
where P_s is electric power for 1 m², W/m²; ΔV is increasing air speed into engine, m/s; m_s is air exemption mass passed throw engine in one second, kg/s·m²; A_s is energy, J/m².

Example 3. Let us take the $U = 10^6$ V, $I_s = 10$ A/m², $\rho = 1$ kg/m³. Then the start thrust is $T_s \approx 7.35 \cdot 10^4$ N/m² = 7.35 tons/m² if the start power is $P_s \approx 10^7$ W/m², exit speed $\Delta V = 270$ m/s.

2. Efficiency of Electron ABEP engine. Efficiency η of any jet (air flight) propulsion is production of two values: propulsion efficiency η_p and engine (cycle) efficiency η_e:

$$\eta = \eta_p \eta_e, \quad \text{where} \quad \eta_p = V/(V + 0.5 \Delta V). \quad (7)$$

The flight efficiency for heat and electronic propulsion are same. They depend only on ΔV. But thermodynamic (cycle) efficiency η_e of the heat engine is low about 25 ÷ 35%. The heat engine looses a great deal of energy from the hot exit jet. For high speed over $M > 3$ the conventional air rocket (jet) engine (TRP) looses efficiency very quickly, because air has dessaciation and ionization at high temperature. The aviation designers try to use the hydrogen fuel, but after $M > 5$ the hydrogen fuel is also useless. The offered electronic jet engine accelerates air by electricity. It has efficiency close to 100% as the only loss of energy is the extraction of the electrons from cathode, ionizations of air molecules and the compensator. This energy is about some electron-volts (eV). The energy spent for acceleration of the air molecules by electrons/ions is hundreds of thousands of eV. That means the total efficiency of ABEP is 3 times more than conventional air jet propulsion.

The second very important point: electric efficiency of ABEP does not depend upon speed of apparatus.

The other advantages: we can make a very large entrance area of engine, we can use the fuselage and wings, stabilizer and keel of plane as engine.

3. Ion and electron speed.

Ion mobility. The ion speed onto the gas (air) jet may be computed by equation:

$$j_s = qn_-b_-E + qD_-(dn_-/dx), \qquad (8)$$

where j_s is density of electric currency about jet, A/m²; $q = 1.6 \times 10^{-19}$ C is charge of single electron, C; n_- is density of injected negative charges in 1 m³; b_- is charge mobility of negative charges, m²/sV; E is electric intensity, V/m; D_- is diffusion coefficient of charges; dn_-/dx is gradient of charges. For our estimation we put $dn_-/dx = 0$. In this case

$$j_s = qn_-b_-E, \quad Q = qn, \quad v = bE, \quad j_s = Qv, \qquad (9)$$

where Q is density of the negative charge in 1 m³; v is speed of the negative charges about jet, m/s.

The air negative charge mobility for normal pressure and temperature $T = 20°C$ is:

In dry air $b_- = 1.9 \times 10^{-4}$ m²/sV, in humid air $b_- = 2.1 \times 10^{-4}$ m²/sV. $\qquad (10)$

In Table 1 there is given the ions mobility of different gases for pressure 700 mm Hg and for $T = 18$ °C.

Table 1. Ions mobility of different gases for pressure 700 mm Hg and for $T = 18$ °C.

Gas	Ion mobility 10^{-4} m²/sV, b_+, b_-		Gas	Ion mobility 10^{-4} m²/sV, b_+, b_-		Gas	Ion mobility 10^{-4} m²/sV, b_+, b_-	
Hydrogen	5.91	8.26	Nitrogen	1.27	1.82	Chloride	0.65	0.51
Oxygen	1.29	1.81	CO_2	1.10	1.14			

Source [22] p.357.

In diapason of pressure from 13 to 6×10^6 Pa the mobility follows the Law bp = const, where p is air pressure. When air density decreases, the charge mobility increases. The mobility strength depends upon the purity of gas. The ion gas mobility may be recalculated in other gas pressure p and temperature T by equation:

$$b = b_0 \frac{T p_0}{T_0 p},\tag{11}$$

where lower index "$_0$" mean the initial (known) point. At the Earth surface $H = 0$ km, $T_0 = 288$ K, $p = 1$ atm; at altitude $H = 10$ km, $T_0 = 223$ K, $p = 0.261$ atm;

For normal air density the electric intensity must be less than 3 MV ($E < 3$ MV/m) and depends from pressure.

Electron mobility. The ratio $E/p \approx$ constant. Conductivity σ of gas depends upon density of charges particles n and their mobility b, for example:

$$\sigma = neb, \quad \lambda = 1/n\sigma,\tag{12}$$

where b is mobility of the electron, λ is a free path of electron.

Electron mobility depends from ratio E/n. This ratio is given in Table 2.

Table 2. Electron mobility b_e in gas vs E/n

Gas	$E/n \times 10^{-17}$ 0.03 V·cm²	$E/n \times 10^{-17}$ 1 V·cm²	$E/n \times 10^{-17}$ 100 V·cm²	Gas	$E/n \times 10^{-17}$ 0.03 V·cm²	$E/n \times 10^{-17}$ 1 V·cm²	$E/n \times 10^{-17}$ 100 V·cm²
N_2	13600	670	370	He	8700	930	1030
O_2	32000	1150	590	Ne	16000	1400	960
CO_2	670	780	480	Ar	14800	410	270
H_2	5700	700	470	Xe	1980	-	240

Source: Physic Encyclopedia http://www.femto.com.ua/articles/part_2/2926.html

The electrons may connect to the neutral molecules and produce the negative ions (for example, affinity of electron to O_2 equals $0.3 \div 0.87$ eV [21] p.424). That way the computation the mobility of a gas contains the electrons and ions is complex problem. Usually the computations are made for all electrons converted to ions.

If $v > 0$, the charged particles accelerate the air into engine ($E > 0$ and engine spend energy). If $v < 0$, the charged particles beak the air into engine ($E < 0$ and engine can produce energy). If $v = 0$ (charged speed about apparatus equals 0), the electric resistance of jet into engine is zero.

The maximal electric intensity in air at the Earth surface is $E_m = 3$ MV/m. If atmospheric pressure changes the E_m also changes by law $E_m/p =$ constant.

Example 4. If $E = 10^6$ V/m, than $v = 200$ m/s in the Earth surface conditions.

4. Electron injectors.

There are some methods for getting the electron emissions: hot cathode emission, cold field electron emission (edge cold emission, edge cathode). The photo emission, radiation emission, radioisotope emission and so on usually produce the positive and negative ions together. We consider only the hot emission and the cold field electron emission (edge cathodes).

Hot electron emission.

Currency i of diode from potential (voltage) U is

$$i = CU^{3/2} \qquad (13)$$

where C is constant which depends from form and size cathode. For plate diode

$$C = \frac{4}{9}\varepsilon_0 \frac{S}{d^2}\sqrt{\frac{2e}{m_e}} \approx 2.33 \cdot 10^{-6} \frac{S}{d^2}, \qquad (14)$$

where $\varepsilon_o = 8.85 \cdot 10^{-12}$ F/m; S is area of cathode (equals area of anode), cm²; d is distance between cathode and anode, cm; e/m_e is the ratio of the electron charge to electron mass, C/kg;

Result of computation equation (13) is in fig.4.

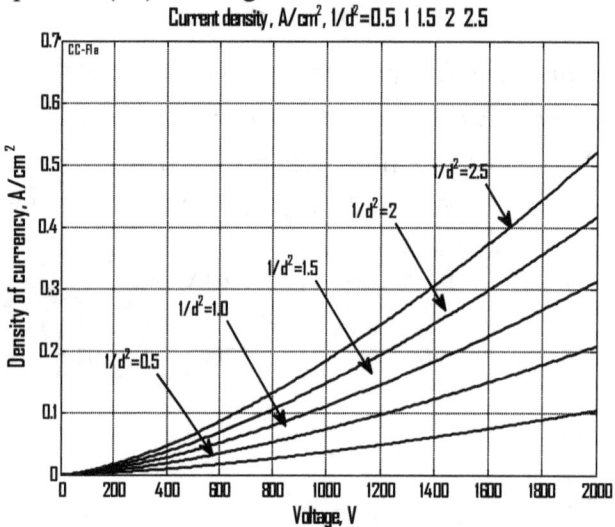

Fig.4. Electric currency via voltage the plain cathodes for different ratio of the distance.

The maximal **hot cathode** emission computed by equation:

$$j_s = BT^2 exp(-A/kT), \qquad (15)$$

where B is coefficient, A/cm²K²; T is catode temperature, K; $k = 1.38 \times 10^{-23}$ [J/K] is Bolzmann constant; $A = e\varphi$ is thermoelectron exit work, J ; φ is the exit work (output energy of electron) in eV, $e = 1.6 \cdot 10^{-19}$. Both values A, B depend from material of cathode and its cover. The "A" changes from 1.3 to 5 eV, the "B" changes from 0.5 to 120 A/cm²K². Boron thermo-cathode produces electric currency up 200 A/cm². For temperature 1400 ÷1500K the cathode can produce currency up 1000 A/cm². The life of cathode can reach some years [20]-[21].

Exit energy from metal are (eV):

$$\text{W 4.5, Mo 4.3, Fe 4.3, Na 2.2 eV}, \qquad (16)$$

From cathode covered by optimal layer(s) the exit work is in Table 3.

Table 3. Exit work (eV) from cathode is covered by the optimal layer(s):

Cr – Cs	Ti – Cs	Ni – Cs	Mo – Cs	W – Ba	Pt – Cs	W – O – K	Steel- Cs	Mo$_2$C-Cs	WSi$_2$-Cs
1.71	1.32	1.37	1.54	1.75	1.38	1.76	1.52	1.45	1.47

Source [20]: Kikoin, Table of physic values, 1976, p. 445 (in Russian).

Results of computation the maximal electric currency (in vacuum) via cathode temperature for the different exit work of electrons *f* are presented in fig.5.

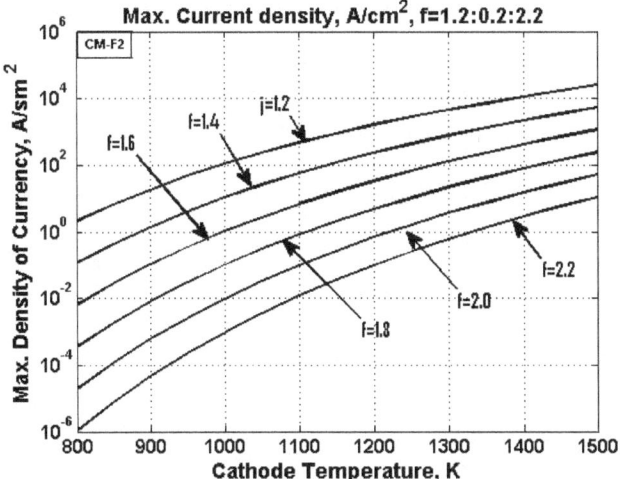

Fig.5. The maximal electric currency via cathode temperature for the different exit work of electrons *f*.

Method of producing electrons and positive ions is well developed in the ionic thrusters for space apparatus.

The field electron emission

  The edge cold emission. The cold field electron emission uses the edge cathodes. It is known that the electric intensity E_e in the edge (needle) is

$$E_e = U/a . \tag{17}$$

Here *a* is radios of the edge. If voltage between the edge and nears net (anode) is $U = 1000$ V, the radius of edge $a = 10^{-5}$ m, electric intensity at edge is the $E_a = 10^8$ V/m. That is enough for the electron emission. The density of electric current may reach up 10^4 A/cm^2. For getting the required currency we make the need number of edges.

The density of electric currency approximately is computed by equation:

$$j \approx 1.4 \cdot 10^{-6} \frac{E^2}{\varphi} 10^{(4.39\varphi^{-1/2} - 2.82 \cdot 10^7 \varphi^{3/2}/E)}, \tag{18}$$

where *j* is density of electric currency, A/cm^2; *E* is electric intensity near edge, V/cm; φ is exit work (output energy of electron, field electron emission), eV.

The density of currency is computed by equation (18) in Table 4 below.

$\varphi = 2.0$ eV		$\varphi = 4.5$ eV		$\varphi = 6.3$ eV	
$E \times 10^{-7}$	lg j	$E \times 10^{-7}$	lg j	$E \times 10^{-7}$	lg j
1,0	2,98	2,0	-3,33	2,0	-12,9
1,2	4,45	3,0	1,57	4,0	-0,88
1,4	5, 49	4,0	4,06	6,0	3,25
1,6	6,27	5,0	5,59	8,0	5,34
1,8	6,89	6,0	6,62	10,0	6,66
2,0	7,40	7,0	7,36	12,0	7,52
2,2	7,82	8,0	7,94	14,0	8,16
2,4	8,16	9,0	8,39	16,0	8,65
2,6	8,45	10.0	8,76	18,0	9,04

Example: Assume we have needle with edge $S_1 = 10^{-4}$ cm², $\varphi = 2$ eV and net $S_2 = 10 \times 10 = 10^2$ cm² located at distance $L = 10$ cm. The local voltage between the needle and net is $U = 10^2$ volts. Than electric intensity at edge of needle, current density and the electric currency is:

$$E = \frac{S_2 U}{S_1 L} = \frac{10^2 10^2}{10^{-4} 10^1} = 10^7 \text{ V/cm}, \quad j = 10^3 \text{ A/cm}^2, \quad i = jS_1 = 10^3 10^{-4} = 0.1 \text{ A}, \tag{19}$$

Here j is taken from Table 4 or computed by equation (18). If we need in the electric currency 10 A, we must locate 100 needles in the entrance area 1×1 m of engine.

Computation of equation (18) is presented in fig. 6.

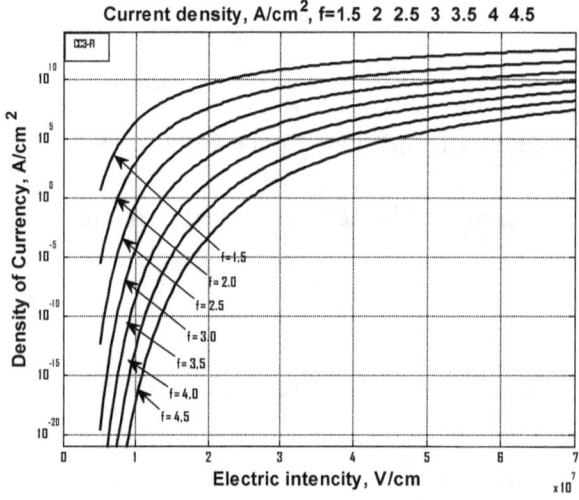

Fig.6. Density of electric currency the noodle injector via the electric intensity for different the field electron emissions f.

Internal and outer pressure on the engine surface.

The electric charges located in the ABEP engine produce electric intensity and internal and outer pressure. The electric intensity can create the electrical breakdown; the pressure can destroy the engine.
a) For the cylindrical engine the electric intensity and pressure may be estimated by equations:

$$E = k\frac{2\tau}{\varepsilon r}, \quad \tau = \frac{i}{V_a}, \quad V_a \approx V + 0.5 \cdot \Delta V, \quad p = E\sigma, \tag{20}$$

where E is electric intensify, V/m; $k = 9 \cdot 10^9$ is electric constant, Nm²/C²; τ is the linear charge, C/m; ε is dielectric constant for given material ($\varepsilon = 1 \div 1000$), r is radius of engine. m; i is electric currency A; V_a is average speed of flow inside of engine, m/s; p is pressure, N/m²; σ is the density of charge, C/m² at an engine surface.

Example. Assume the engine has $r = 0.5$ m, $V = 270$ m/s, $\Delta V = 200$ m/s, $i = 5$ A. Let us take as isolator the Lexan having the dielectric strength $E_m = 640$ MV/m and $\varepsilon = 3$. Than from (14) we have $E = 81$ MV/m $< E_m = 640$ MV/m.

If $E > E_m$ we can locate the part of the compensate charge inside engine.

b) For plate engine having entrance h×w = 1× 3 m and compensation charges on two sides, the electric intensity and pressure may be estimated by equations:

$$E = 4\pi k \frac{\sigma}{\varepsilon}, \quad \sigma = \frac{i}{2V_a w}, \quad V_a \approx V + 0.5 \cdot \Delta V, \quad p = 2\varepsilon\varepsilon_0 E^2, \qquad (21)$$

where w is width of entrance, m; ε is dielectric coefficient of the isolator.

5. Electrical Generator

Suggested engine needs a great deal of electricity which can be gotten either from a nuclear reactor or from connection of the conventional turbojet engine with an electric generator. Let us consider the last possibility.

When aircraft is in needs of electricity, most aviation engineers offer the conventional way: take the usual magnetic electric generator and connect it to the turbojet or take other (for example, piston) engine. Let us analyze the limiting possibilities of different versions.

Magnetic electric generator. Magnetic electric generator was first produced about century ago and has been very well studied. The ratio of power/mass of magmatic generator for 1 m³ may be estimated by equation:

$$A = \frac{B^2}{2\mu_0}, \quad P = Av, \quad M = \gamma, \quad \frac{P}{M} = c\frac{B^2 v}{2\mu_0 \gamma}, \qquad (22)$$

Here A is density of energy into 1 m³ of magnetic material J/m³; B is maximal magnetic intensity, T; $\mu_0 = 4\pi 10^{-7}$ is permeability (magnetic constant), N/A²; P is power, W; v is electric frequency, 1/s ($v = 50 \div 400$ 1/s); M is mass 1 m³ of generator, kg/m³; γ is specific mass of the generator bogy, kg/m³ ($\gamma \approx 8000$ kg/m³); $c \approx 1/8$ correction coefficient, because average $B = 0.5 B_{max}$ and ferromagnetic iron uses only about ½ engine volume. The maximal frequency determinates the ratio L/r, where L is inductance, r is electric resistance. That equals about 500 – 1000 1/s.

Example. Let us take the typical data $B = 1$ T, $v = 400$ 1/s, $\gamma = 8000$ kg/m³. We get maximal $P/M = 2.5$ kW/kg.

Typical aviation generator has:

Type: ГТ-120ПЧ8 (Russian)	Power	Phases	Voltage	Currency	Frequency	Number of rev.	Mass
	120 kW	3	208V	334 A	400 1/s	8000 in min	90 kg

The ratio for the usual aircraft generator equals 1.33 kW/kg. That is two time less than maximal possible. For our purposes that will be two times less because we need high voltage. But the high voltage transformer will weigh not less than electric generator. If aircraft has turbo 10,000 kW the magnetic propulsion system will weigh about 14 tons, 5 times more than turbojet. That is not acceptable in aviation. In addition, we need a constant (direct) current. The generator of DC weighs significantly more. No suitable transformer for transformation the DC into in very high voltage.

Electrostatic generator (EG). Electrostatic electric generator is known for about two centuries but it is not used because it produces very high voltage which is very dangerous for people and not suitable for practice and home devises. As a result, EG is studied very little and no power EG is produced by industry.

The ratio power/mass of electrostatic generator for area $S = 1$ m² may be estimated by equation:

$$A = \frac{CU^2}{2}, \quad C = \frac{\varepsilon\varepsilon_0 S}{d}, \quad U = Ed, \quad P = \frac{A}{t}, \quad M = \gamma\delta S, \quad \frac{P}{M} = \frac{\varepsilon\varepsilon_0 E^2 \delta}{2\gamma d^2} V_a, \tag{23}$$

where A is density of energy on 1 m² of the electrostatic (isolator plate) material J/m²; C is capacitance of plate (one plate of condenser), F/m²; U is voltage, V; ε (1 ÷ 3000) dielectric constant of plate matter; $\varepsilon_0 = 8.85 \cdot 10^{-12}$ is permittivity, F/m; S is area of one plate, m²; d is distance between plates (include thickness of one plate, m); P is power, W; t is time, s; M is mass 1 m² of generator plate, kg/m³; γ is specific mass of the generator plate, kg/m³ ($\gamma \approx 1800$ kg/m³), δ is clearance between plates, m; V_a is the average relative speed of two plates, m/s ($V_a \approx 0.5V$, where V is the peripheral disk (plate) speed).

Properties of some insulators in Table 5.

Table 5. Properties of various good insulators (recalculated in metric system)

Insulator	Resistivity Ohm-m.	Dielectric strength MV/m.. E_i	Dielectric constant, ε	Tensile strength kg/mm², $\sigma \times 10^7$ N/m²
Lexan	10^{17}–10^{19}	320–640	3	5.5
Kapton H	10^{19}–10^{20}	120–320	3	15.2
Kel-F	10^{17}–10^{19}	80–240	2–3	3.45
Mylar	10^{15}–10^{16}	160–640	3	13.8
Parylene	10^{17}–10^{20}	240–400	2–3	6.9
Polyethylene	10^{18}–5×10^{18}	40–680*	2	2.8–4.1
Poly (tetra-fluoraethylene)	10^{15}–5×10^{19}	40–280**	2	2.8–3.5
Air (1 atm, 1 mm gap)	-	4	1	0
Vacuum (1.3×10^{-3} Pa, 1 mm gap)	-	80–120	1	0

*For room temperature 500–700 MV/m.
** 400–500 MV/m.

Source: Encyclopedia of Science & Technology[9] (Vol. 6, p. 104, p. 229, p. 231).(See also [10], p.283.
Note: Dielectric constant ε can reach 4.5 – 7.5 for mica (E is up 200 MV/m); 6 – 10 for glasses (E = 40 MV/m) and 900 – 3000 for special ceramics (marks are CM-1, T-900) [21] p.32 (E = 13 – 28 MV/m). Dielectric strength appreciable depends from surface roughness, thickness, purity, temperature and other conditions of material. It is necessary to find good insulate materials and reach conditions which increase the dielectric strength.

The safety peripheral disk speed may be estimated by equation $V = (\sigma/\gamma)^{0.5}$ where σ is safety tensile stress (N/m²), γ is specific weight, kg/m³. The disk may be reinforced by fiber having high tensile stress.

Let us consider the following experimental and industrial fibers, whiskers, and nanotubes:

1. Experimental nanotubes CNT (carbon nanotubes) have a tensile strength of 200 Giga-Pascals (20,000 kg/mm^2), Young's modulus is over 1 Tera Pascal, specific density $\gamma = 1800$ kg/m^3 (1.8 g/cc) (year 2000). For safety factor $n = 2.4$, $\sigma = 8300$ kg/mm^2 = 8.3×10^{10} N/m^2, $\gamma = 1800$ kg/m^3, $(\sigma/\gamma) = 46 \times 10^6$, $K = 4.6$. The SWNTs nanotubes have a density of 0.8 g/cc, and MWNTs have a density of 1.8 g/cc. Unfortunately, the nanotubes are very expensive at the present time (1994).
2. For whiskers C_D $\sigma = 8000$ kg/mm^2, $\gamma = 3500$ kg/m^3 (1989) [10].
3. For industrial fibers $\sigma = 500 - 600$ kg/mm^2, $\gamma = 1800$ kg/m^3, $\sigma/\gamma = 2{,}78 \times 10^6$, $K = 0.278 - 0.333$,

Figures for some other experimental whiskers and industrial fibers are given in Table 6.

Table 6. Properties of fiber and whiskers

Material Whiskers	Tensile strength kgf/mm^2	Density g/cc	Material Fibers	Tensile strength MPa	Density g/cc
AlB$_{12}$	2650	2.6	QC-8805	6200	1.95
B	2500	2.3	TM9	6000	1.79
B$_4$C	2800	2.5	Thorael	5650	1.81
TiB$_2$	3370	4.5	Allien 1	5800	1.56
SiC	1380–4140	3.22	Allien 2	3000	0.97

See Reference [10] p. 33.

Example: Let us estimate ratio P/M of the electrostatic generator by equation (23). Take the electric intensity $E = 10^7$ V/m, area of the disk 1 m^2, thickness of the disk 0.003m, clearance between disks $\delta = 0{,}002$ m, ($d = 0.005$ m), $V = 500$ m/s, $\gamma = 1800$ kg/m^3, $\varepsilon = 3$. Substitute these data in equation (23) we get $P/M = 53$ kW/kg. That means the electrostatic generator (motor) of equal power will be in 20 times less than magnetic generator (motor). The 10,000 kW electrostatic generator (motor) will be weight only 400 kg (200 disks). And additional the electrostatic generator produces high voltage direct (constant) electric currency. The powerful turbo-propeller jet HK-12 (Russia) has a start power 8700 kW and mass 2800 kg. The propeller (5.6 m) weights 1156 kg in it. We can delete propeller, installs the electrostatic generator (volume 1 m^3), the light offered ABEP engine and flights with hypersonic speed. The electricity easy transverses to other (for example to VTOL) engine.

Air friction in electrostatic generator and its efficiency.

Let us estimate ratio of the air friction/produced power 1 m^2 of disk the electrostatic generator. Compute the friction, produced power and efficiency:

$$P_f = 2FV_a, \quad F = \varsigma \frac{V_a}{\delta}, \quad P_f = 2\frac{\varsigma V_a^2}{\delta}, \quad P = \frac{\varepsilon \varepsilon_0 E^2}{2} V_a, \quad \eta = 1 - \frac{P_f}{P} = 1 - \frac{4\varsigma V_a}{\varepsilon \varepsilon_0 \delta E^2}, \quad (24)$$

where P_f is power of friction 1 m^2 of disk, W/m^2; F is friction force 1 m^2 of disk, N/m^2; V_a is average disk speed, m/s; ς is viscosity of the gas (for air $\varsigma = 1.72 \cdot 10^{-5}$ Pa·s, for hydrogen $\varsigma = 0.84 \cdot 10^{-5}$ at atmospheric pressure and $T = 0°$C); P is power produced 1 m^2 of disk, W/m^2; ε is dielectric constant of plate matter; $\varepsilon_0 = 8{,}85 \cdot 10^{-12}$ is permeability, F/m; δ is clearance between disk, m; E is electric intensity, V/m; η is efficiency of generator related to air friction.

Example: If $V_a = 250$ m/s; $E = 2 \cdot 10^6$ V/m; $\delta = 0.002$ m, then $\eta = 0.92$.

The coefficient of gas friction weak depends from the pressure and temperature. If we change the air into the electrostatic generator by hydrogen, the loss of friction decreases in two times. If we create the vacuum into the electrostatic generator, the gas friction will be zero and the safety electric intensity is increased in many times.

Loss of energy and matter for ionization.

Let us estimate the energy and matter is requested for ionization and discharge the offered ABEP propulsion. Assume we have ABEP engine having the power $P = 10,000$ kW and a work voltage $V = 1$ MV. In this case the electric currency is $i = P/V = 10$ A $= 10$ C/s.

Assume we use the nitrogen N_2 for ionization (very bad gas for it). It has exit work about 5 eV and relative molecular weight 14. One molecule (ion) of N_2 weights $m_N = 14 \cdot 1.67 \cdot 10^{-27} = 2.34 \cdot 10^{-26}$ kg. The 1 ampere has $n_A = 1/e = 1/1.6 \cdot 10^{-19} = 6.25 \cdot 10^{18}$ ions/s. Consumption of the ion mass is:
$M = m_N i \, n_A = 2.34 \cdot 10^{-26} \cdot 10 \cdot 6.25 \cdot 10^{18} = 1.46 \cdot 10^{-6}$ kg/s $= 1.46 \cdot 10^{-6} \cdot 3.6 \cdot 10^{-3} = 5.26 \cdot 10^{-3}$ kg/hour ≈ 5 gram/hour.

If electron exit work equals $\varphi = 4.5$ eV the power spent extraction of one electron is: $E_1 = \varphi e = 4.5 \cdot 1.6 \cdot 10^{-19} = 7.2 \cdot 10^{-19}$ J.

The total power for the electron extraction is $E = i \cdot n_A \cdot E_1 = 10 \cdot 6.25 \cdot 10^{18} \cdot 7.2 \cdot 10^{-19} = 45$ W.

The received values mass M and power E are very small in comparison with conventional consumption of fuel (tons in hour) and engine power (thousands of kW).

Important note (Compensation of flow charge). Any contact collector cannot collect ALL charges. Part of them will fly away. That means the engine (apparatus) will be charged positive (if fly away electrons or negative ions) or negative (if fly away the positive ions). It is easy delete the negative charges by edge. The large positive charge we may delete by small ion accelerator. The ion engines (trasters) for vacuum are R&D well. They may be used as injectors and dischargers in the first design of ABEP.

Fuel efficiency of ABEP engine.

For passenger and transport aircraft about half of the cost of transportation is cost of fuel. Let us estimate the consumption of fuel for ABEP engine in hypersonic flight. The hypersonic flight has three stages [1]: acceleration, ballistic, braking. But in contrast to subsonic plane the hypersonic plane expends fuel only in the first stage: acceleration. The second stage "ballistic" for "flight with constant hypersonic speed in atmosphere", usually requires more fuel than subsonic flight (because the coefficient of aerodynamic efficiency of subsonic flight in 2 – 3 times more than hypersonic flight).

a) Data of the Stage of acceleration the mass 1 kg up the given high hypersonic speed V is

$$L_a = \frac{V^2}{2a}, \quad E_a = \frac{V^2}{2}, \quad E_d = F \cdot L_a = \frac{g}{K_2} \cdot \frac{V^2}{2a}, \quad E = E_a + E_d = \frac{V^2}{2}\left(1 + \frac{g}{aK_2}\right), \quad (25)$$

where L_a is distance of acceleration, m; V is final speed of acceleration, m/s; $a > 0$ is acceleration, m/s²; E_a is energy of acceleration 1 kg, J; E_d is energy of drag (1kg) in acceleration distance, J; E is full energy (for mass 1 kg) spent in the acceleration distance, J; $g = 9.81$ m/s² is gravitation; K_2 is ratio Lift/Drag (coefficient of aerodynamic efficiency, for $M > 1.5$, $K_2 \approx 4(1 + 1/M)$, where M is Max number).

b) Flight in ballistic trajectory at high altitude in a rare atmosphere with neglect the air drag, $3 < M <$

6 is described:

$$L_b = \frac{V^2 \sin 2\alpha}{g}, \quad H_b = \frac{V^2 \sin^2 \alpha}{2g}, \quad (26)$$

where L_b is distance of ballistic trajectory, m; H_b is maximal altitude, m; α is angle of incidence into rare atmosphere (optimal $\alpha \approx 45°$).

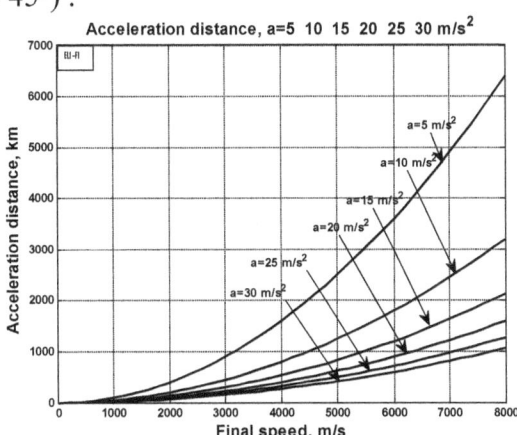

Fig. 7. Acceleration distance of hypersonic aircraft via final speed of ABEP for different acceleration m/s².

c) Flight in high ballistic trajectory in high altitude at vacuum with neglect the air drag, M > 10, (the angle of incidence is optimal ($\alpha \approx 30° - 40°$)) is described.

$$v = \left(\frac{V}{V_0}\right)^2, \quad v < 1, \quad tg\beta = \frac{v}{2\sqrt{1-v}}, \quad L_w = 2R\beta, \quad (27)$$

where v is relative speed; V_0 = 7.93 km/s is circle space speed of Earth satellite, m/s; β is angle from entrance in space to maximal altitude measured from Earth center, rad; R = 6378 km – radius of Earth, km; L_w is distance ballistic flight into space, km.

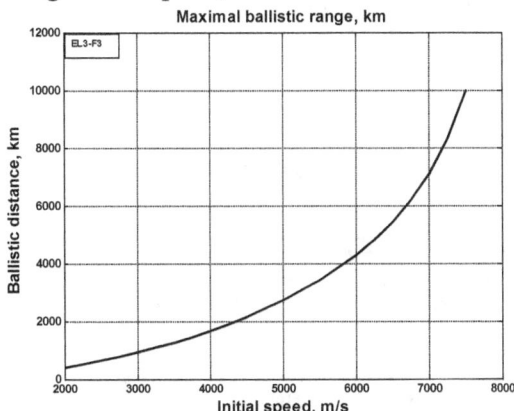

Fig.8. maximal ballistic range via initial speed of space sraft.

d) Brake (inertial, kinetic) flight into atmosphere [3].

$$L_i \approx \frac{K_2 V^2}{2g}, \quad (28)$$

e) Total range of hypersonic aircraft is (with exit to space, $M > 10$)

$$L = L_a + L_w + L_i. \quad (29)$$

If $M < 10$ and there is exit to ballistic trajectory, the full range is
$$L = L_a + L_b + L_i . \tag{30}$$
If $M < 10$ and **no** exit to ballistic trajectory, the range is
$$L = L_a + L_h + L_i, \text{ where } L_h = E_h K_2 / g , \tag{31}$$
where L_h is distance with constant hypersonic flight, m or km; E_h is energy spent in this distance. where K_2 is coefficient of aerodynamic efficiency (ratio Lift/Drag).

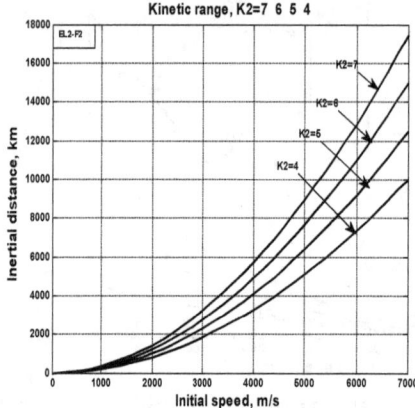

Fig.9. Kinetic (inertial) range of hypersonic aircraft via initial speed.

The total range compare with range of the conventional subsonic aircraft:
$$L_s \approx \frac{E K_1}{g}, \text{ where } E = E_a \text{ or } E = E_a + E_h , \tag{32}$$
Here is L_s is range of the subsonic aircraft, m or km; K_1 is ratio Lift/Drag (coefficient of aerodynamic efficiency of subsonic aircraft), $K_1 \approx 10 \div 18$.

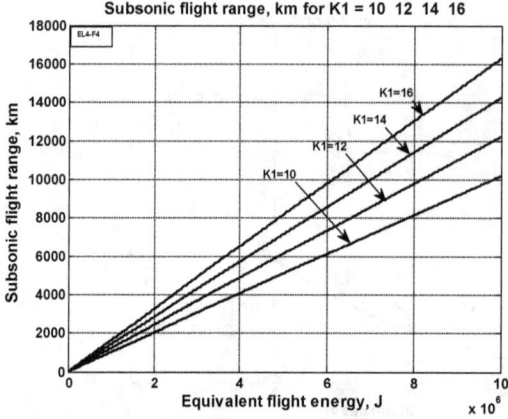

Fig.10. Flight range of subsonic aircraft via the equivalent flight energy for different ratio Lift/Drag.

If range L_s of the subsonic aircraft is less than range of hypersonic aircraft, the hypersonic aircraft is more profitable (spend less fuel in 1 kg·km) than subsonic plane, and conversely (without accounting of other advantages the hypersonic speed!).

The estimations show: in speed diapason M ≈ 1,5 ÷ 5 the supersonic aircraft spend fuel in 1.5 – 2.5 more than subsonic airplane, but after M > (12 ÷ 20) hypersonic aircraft with ABEP engine spent less the fuel. If M > 25 the range may be any for constant fuel consumption. The men or load can be delivered in any point of the Earth in during 45 – 50 min. That means the cost of travel through space may be faster and cheaper than long distance travel by subsonic airplane. For example, the time of flight are: New-York – Paris 12 min (5837 km), San-Francisco - Tokyo 16 min (8277 km), NY – Moscow 17 min (7519 km).

The offered engine as rocket motor.

Offered engine has the principal differences from rocket engine, in particular: one need in environment (which it accelerates) and electric energy. Unlike rockets and most space propulsions methods, ABEP engine does not use a thermal principle but rather accelerates an environment matter by electrons. But if ABEP has as source of electric energy such as a nuclear reactor, the offered engine may be used as very high efficiency rocket engine.

Conventional nuclear rocket engine has limited impulse (limited exhaust speed of gas) because engine material has limited temperature and gas became dissociate. Rockets need **special fuel**. The ABEP engine is limited only by electric power (energy). It can have a higher impulse and (main advantage) can use ANY matter as propulsion material. ABEP having nuclear electric reactor can use (as refuel) any space body (meteorites, asteroids, planets) to refill its fuel supply. If astronauts (being on outer body) have a choice between using a fuel for conventional rocket and by ABEP may be situation when a using ABEP is more useful.

Let us make the estimation of the next case. Space ship having total mass $M_o = 1000$ kg has $M_F = 100$ kg fuel (oxygen – hydrogen) and must start from the asteroid with maximal speed. Ship can use conventional rocket engine or take 9000 kg of asteroid matter (in dust form), convert fuel in electricity and use offered engine. Equations requested for estimation are below:

$$V_1 = -w_1 \ln \frac{M_f}{M_0}, \quad E = \eta M_F E_1, \quad w_2 = \left(\frac{2E}{M_{0,a}}\right)^{0.5}, \quad V_2 = -w_2 \ln \frac{M_{f,a}}{M_{0,a}}, \quad (33)$$

where $w_1 = 4000$ m/s is speed of rocket exhaust gas (rocket impulse in m/s); $M_f = 900$ kg is final ship mass; E is fuel energy, J; $\eta \approx 0.5$ is total coefficient convert efficiency; $E_1 \approx 13.45 \cdot 10^6$ J is energy ability of fuel (oxygen – hydrogen) ; $w_2 \approx 348$ m/s is computed impulse ABEP in given case; $V_2 \approx 800$ m/s is new speed of space ship with offered propulsion.

Using rocket theory (33) we find: ship can reach the speed of about 400 m/s by a conventional rocket engine and 800 m/s by the offered propulsion system. Than means two times more than conventional methods.

SUMMARY AND DISCUSSION.

The author proposed a fundamentally new propulsion system (engine) using the environment medium (air, space material) and electric energy. It is not comparable to conventional heat propulsion because the usual heat jet engine gets the thrust by compressing the air, burning the fuel into air, heating of air, accelerating the hot air and expiring the hot gas in atmosphere.

The offered ABEP engine is accelerating the air (medium) by a principally new method – by electrons and electric field which does not need atmospheric oxygen and thus can work in any atmosphere of other planets. This engine does not require compressing and heating of medium and, as such, does not have limitations of high temperature, high flight speed and composition of the atmosphere.

This engine is also dissimilar to the known space electric engines. The conventional space electric engine takes an extracted mass from itself, ionizes it, and accelerates springing forward in a vacuum. It has very small thrust, works poorly into any atmosphere and works worse if the atmosphere has a high density. The ABEP does not take the extracted mass, can work only in atmosphere and works better if the atmosphere has a high density.

The main disadvantage of the offered engine is the requirement of high voltage electricity. For getting the electricity it may use the conventional internal turbo engine connected to electrostatic generator. Electrostatic power generator is lightweight and produces high voltage electricity. The weight system turbojet engine + electrostatic generator + is same (or less) than mass system turbojet engine + propeller system.

Researches related to this topic are presented in [1]-[19]. See also [20]-[24].

ACKNOWLEDGEMENT

The author wishes to acknowledge Shmuel Neumann for correcting the English and offering useful advice and suggestions.

References

[1]. A.A. Bolonkin, Electron Air Hypersonic Propulsion. International Journal of Advanced Engineering Applications, Vol.1, Iss.6, pp.42-47 (2012). http://viXra.org/abs/1306.0003, http://www.scribd.com/doc/145165015/Electron-Air-Hypersonic-Propulsion , http://www.scribd.com/doc/146179116/Electronic-Air-Hypersonic-Propulsion , http://fragrancejournals.com/wp-content/uploads/2013/03/IJAEA-1-6-6.pdf

[2] A.A. Bolonkin, Air Catapult Transportation. NY, USA, Scribd, 2011.
Journal of Intelligent Transportation and Urban Planning (JTUP), April 2014, Vol.2, pp. 70-84.
http://www.scribd.com/doc/79396121/Article-Air-Catapult-Transportation-for-Scribd-1-25-12, http://www.archive.org/details/AirCatapultTransport, http://viXra.org/abs/1310.0065 .
Chapter in Book: Recent Patents on Electrical & Electronic Engineering, Bentham Science Publishers, Vol.5, No.3, 2012.

[3]. A.A. Bolonkin, "High Speed Catapult Aviation", AIAA-2005-6221, presented to *Atmospheric Flight Mechanic Conference* – 2005. 15–18 August, USA.

[4]. A.A. Bolonkin, "Air Cable Transport System", *Journal of Aircraft*, Vol. 40, No. 2, July-August 2003, pp. 265–269.

[5]. A.A. Bolonkin, "Bolonkin's Method Movement of Vehicles and Installation for It", US Patent 6, 494, 143 B1, Priority is on 28 June 2001.

[6]. A.A. Bolonkin, "Air Cable Transport and Bridges", TN 7567, *International Air & Space Symposium* – The Next 100 Years, 14-17 July 2002, Dayton, Ohio, USA

[7]. A.A. Bolonkin, "Non-Rocket Missile Rope Launcher", IAC-02-IAA.S.P.14, *53rd International Astronautical Congress, The World Space Congress – 2002*, 10–19 Oct 2002, Houston, Texas, USA.

[8]. A.A. Bolonkin, "Inexpensive Cable Space Launcher of High Capability", IAC-02-V.P.07, *53rd International Astronautical Congress. The World Space Congress – 2002*, 10–19 Oct. 2002. Houston, Texas, USA.

[9]. A.A. Bolonkin, "Non-Rocket Space Rope Launcher for People", IAC-02-V.P.06, *53rd International Astronautical Congress. The World Space Congress – 2002*, 10–19 Oct 2002, Houston, Texas, USA.

[10]. A.A. Bolonkin, "*Non-Rocket Space Launch and Flight*", Elsevier, 2005, 468 pgs. Attachment 2: High speed catapult aviation, pp.359-369. http://www.archive.org/details/Non-rocketSpaceLaunchAndFlight , http://www.scribd.com/doc/24056182

[11]. A.A. Bolonkin, "*New Concepts, Ideas, Innovations in Aerospace, Technology and the Human Sciences*", NOVA, 2006, 510 pgs. http://www.scribd.com/doc/24057071 , http://www.archive.org/details/NewConceptsIfeasAndInnovationsInAerospaceTechnologyAndHumanSciences

[12]. A.A. Bolonkin, R. Cathcart, "*Macro-Projects: Environments and Technologies*", NOVA, 2007, 536 pgs. http://www.scribd.com/doc/24057930; http://www.archive.org/details/Macro-projectsEnvironmentsAndTechnologies .

[13]. A.A. Bolonkin, *Femtotechnologies and Revolutionary Projects*. Lambert, USA, 2011. 538 p., 16 Mb. http://www.scribd.com/doc/75519828/ , http://www.archive.org/details/FemtotechnologiesAndRevolutionaryProjects

[14]. A.A. Bolonkin, *LIFE. SCIENCE. FUTURE* (Biography notes, researches and innovations). Scribd, 2010, 208 pgs. 16 Mb. http://www.scribd.com/doc/48229884, http://www.archive.org/details/Life.Science.Future.biographyNotesResearchesAndInnovations

[15]. A.A.Bolonkin, "Magnetic Space Launcher" has been published online 15 December 2010, **in the** ASCE, *Journal of Aerospace Engineering* (Vol.24, No.1, 2011, pp.124-134). http://www.scribd.com/doc/24051286/

[16]. A.A.Bolonkin, Universe. Relations between Time, Matter, Volume, Distance, and Energy (part 1) http://viXra.org/abs/1207.0075, http://www.scribd.com/doc/100541327/ , http://archive.org/details/Universe.RelationsBetweenTimeMatterVolumeDistanceAndEnergy

[17]. A.A.Bolonkin, Lower Current and Plasma Magnetic Railguns. Internet, 2008. http://www.scribd.com/doc/31090728 ; http://Bolonkin.narod.ru/p65.htm .

[18] A.A.Bolonkin, Electrostatic Climber for Space Elevator and Launcher. Paper AIAA-2007-5838 for *43 Joint Propulsion Conference*. Cincinnati, Ohio, USA, 9 – 11 July, 2007. See also [12], Ch.4, pp. 65-82.

[19]. A.A. Bolonkin, "Air Cable Transport System", *Journal of Aircraft*, Vol. 40, No. 2, July-August 2003, pp. 265–269.

[20] N.I. Koshkin and M.G. Shirkebich, Directory of Elementary Physics, Nauka, Moscow, 1982 (in Russian).

[21] I.K. Kikoin. Table of Physics values. Atomisdat, Moscow, 1976 (in Russian).

[22] S.G. Kalashnikov, Electricity, Moscow, Nauka, 1985.(in Russian).

[23] W.J. Hesse and el. Jet Propulsion for Aerospace Application, Second Edition, Pitman Publishing Corp. NY.

[24] Wikipedia. Ion craft, http://wikipedia.org .

May 27, 2014

Chapter 9

Electron Hydro Electric Generator

Abstract

Author offers a new method of getting electric energy from moving water. A special injector injects electrons into water. Water stream picks up the electrons and moves them in the direction of stream which is against the direction of electric field. At some distance from injector a unique grid acquires the electrons, thus charging and producing electricity. This method does not require, as does other water energy devices, strong dams, water turbines, or electric generators. The proposed water installation is very cheap. The area of water braking may be large and produces a great deal of energy. This electron water installation may be in river or ocean stream (as Gulf Stream).

Keywords: *water energy, utilization of water energy, electronic water electric generator, WABG, Bolonkin.*

Introduction
Water energy

Hydropower or **water power** is power derived from the energy of falling water and running water, which may be harnessed for useful purposes. Since ancient times, hydropower has been used for irrigation and the operation of various mechanical devices, such as watermills. Since the early 20th century, the term is used almost exclusively in conjunction with the modern development of hydro-electric power, which allowed use of distant energy sources. Hydro power is a renewable energy source.

Rivers. Volumetric flow rate, also known as discharge, volume flow rate, and rate of water flow, is the volume of water which passes through a given cross-section of the river channel per unit time. It is typically measured in cubic meters per second (cumec) or cubic feet per second (cfs), where 1 m^3/s = 35.51 ft^3/s; it is sometimes also measured in liters or gallons per second. Volumetric flow rate can be thought of as the mean velocity of the flow through a given cross-section, times that cross-sectional area. Mean velocity can be approximated through the use of the Law of the Wall. In general, velocity increases with the depth (or hydraulic radius) and slope of the river channel, while the cross-sectional area scales with the depth and the width: the double-counting of depth shows the importance of this variable in determining the discharge through the channel.

Data of Some World Rivers.

Amazon: elevation 5170 m (16,962 ft), length 7,000 km (4300 mi), average discharge 209,000 m^3/s (7,381,000 cu ft/s).

Mississippi: elevation 450 m (1,475 ft), length 3,734 km (2320 mi), average discharge 16,792 m^3/s (593,000 cu ft/s). The Mississippi River discharges at an annual average rate of between 200 and 700 thousand cubic feet per second (7,000–20,000 m^3/s). Although it is the 5th largest river in the world by volume, this flow is a mere fraction of the output of the Amazon, which moves nearly 7 million cubic feet per second (200,000 m^3/s) during wet seasons. On average, the Mississippi has only 8% the flow of the Amazon River.

Niagara Falls is the collective name for three waterfalls that straddle the international border between the Canadian province of Ontario and the U.S. state of New York. They form the southern end of the

Niagara Gorge. Located on the Niagara River, which drains Lake Erie into Lake Ontario, the combined falls form the highest flow rate of any waterfall in the world, with a vertical drop of more than 165 feet (50 m). Horseshoe Falls is the most powerful waterfall in North America, as measured by vertical height and also by flow rate. While not exceptionally high, the Niagara Falls are very wide. More than six million cubic feet (168,000 m^3) of waterfalls over the crest line every minute in high flow, and almost four million cubic feet (110,000 m^3) on average.

Marine energy (also sometimes referred to as **ocean energy** or **ocean power**) also refers to the energy carried by ocean waves, tides, salinity, and ocean temperature differences. The movement of water in the world's oceans creates a vast store of kinetic energy, or energy in motion. This energy can be harnessed to generate electricity to power homes, transport and industries. The term marine energy encompasses both wave power — power from surface waves, and tidal power — obtained from the kinetic energy of large bodies of moving water. The oceans have a tremendous amount of energy and are close to many if not most concentrated populations. Ocean energy has the potential of providing a substantial amount of new renewable energy around the world.

Marine current power is a form of marine energy obtained from harnessing of the kinetic energy of marine currents, such as the Gulf Stream. Although not widely used at present, marine current power has an important potential for future electricity generation. Marine currents are more predictable than wind and solar power.

A 2006 report from United States Department of the Interior estimates that capturing just $1/_{1,000\text{th}}$ of the available energy from the Gulf Stream, which has 21,000 times more energy than Niagara Falls in a flow of water that is 50 times the total flow of all the world's freshwater rivers, would supply Florida with 35% of its electrical needs.

Marine currents are caused mainly by the rise and fall of the tides resulting from the gravitational interactions between earth, moon, and sun, causing the whole sea to flow. Other effects such as regional differences in temperature and salinity and the Coriolis Effect due to the rotation of the earth are also major influences. The kinetic energy of marine currents can be converted in much the same way that a wind turbine extracts energy from the wind, using various types of open-flow rotors. The potential of electric power generation from marine tidal currents is enormous. There are several factors that make electricity generation from marine currents very appealing when compared to other renewables:

- The high load factors resulting from the fluid properties. The predictability of the resource, so that, unlike most of other renewables, the future availability of energy can be known and planned for.
- The potentially large resource that can be exploited with little environmental impact, thereby offering one of the least damaging methods for large-scale electricity generation.
- The feasibility of marine-current power installations to provide also base grid power, especially if two or more separate arrays with offset peak-flow periods are interconnected.

Gulf Stream. As a consequence, the resulting Gulf Stream is a strong ocean current. It transports water at a rate of 30 million cubic meters per second (30 sverdrups) through the Florida Straits. As it passes south of Newfoundland, this rate increases to 150 million cubic meters per second. The volume of the Gulf Stream dwarfs all rivers that empty into the Atlantic combined, which barely total 0.6 million cubic meters per second. It is weaker, however, than the Antarctic Circumpolar Current. The Gulf Stream is typically 100 kilometers (62 mi) wide and 800 meters (2,600 ft.) to 1,200 meters (3,900 ft.) deep. The current velocity is fastest near the surface, with the maximum speed typically about 2.5 meters per second (5.6 mph).

Tidal power, also called **tidal energy**, is a form of hydropower that converts the energy of tides into useful forms of power - mainly electricity. Although not yet widely used, tidal power has potential for future electricity generation. Tides are more predictable than wind energy and solar power. Among sources of renewable energy, tidal power has traditionally suffered from relatively high cost and limited availability of sites with sufficiently high tidal ranges or flow velocities, thus constricting its total availability.

Wave energy is the transport of energy by ocean surface waves, and the capture of that energy to do useful work – for example, electricity generation, water desalination, or the pumping of water (into reservoirs). Machinery able to exploit wave power is generally known as a **wave energy converter** (WEC). Wave power is distinct from the diurnal flux of tidal power and the steady gyre of ocean currents. Wave-power generation is not currently a widely employed commercial technology, although there have been attempts to use it since at least 1890. In 2008, the first experimental wave farm was opened in Portugal, at the Aguçadoura Wave Park.

The realistically usable worldwide resource has been estimated to be greater than 2 TW. Locations with the most potential for wave power include the western seaboard of Europe, the northern coast of the UK, and the Pacific coastlines of North and South America, Southern Africa, Australia, and New Zealand. The north and south temperate zones have the best sites for capturing wave power. The prevailing westerly's in these zones blow strongest in winter.

The reader can find the authors ideas about various innovations in harnessing wind energy in [1]-[7], and additional information about water energy in [8]-[11].

Description of Innovation

Design. One simplest version of the offered electron water generator (WABG) is presented in fig.1. Installation contains: electron injectors 2 established in column 6 and electron collector having the high voltage ring 8 and inside net 4. Net 4 are having the conductive leaves 5 (metallic foil, for example, aluminum foil). They have a large surface which helps to collect the electrons from a large area and send to the ring 8. Network connects with the electron injectors through a useful load 7.

Work of WABG. The WABG generator works the following way: injector injects the electrons into water, the water stream catches them and moves to network 4 of the electron collector 8. Ring 8 has negative charge, electron injector has positive charge. The electric field of ring 8 breaks the electrons (negative ions) and decreases the water speed. But the electric ion speed is significantly less than water stream speed and electrons when they reach the net of collector settle into net 4 and move to ring 8 of the collector and increase its negative charge of the ring 8. Those additional charges (electrons) return through the electric load 7 and make the useful work.

If river is navigable, the collector is located on the river bottom (fig.2). The injectors may be up on a mast (fig. 2a) or located also on river bottom (fig. 2b). The efficiency of these will be different. The surface collector is conductivity film 11 (fig.2) (for example, aluminum foil) which pass them to isolated high voltage ring 8. For increasing the efficiency of collector we can (optionally) place under net of collector the isolated positive charge 12 (or positive electrets) (fig. 2).

Advantages of the proposed electron wind systems (WABG) in comparison with the conventional hydropower systems.

The suggested new principle electron water generator (WABG) has the following advantages in comparison with conventional hydro dam systems used at present time.

Advantages:

1. Offered installations are very simple.
2. Offered system is **very cheap** (by hundreds of times). No dam, hydro-turbines, electric generators, special canals for ships, fish, filling the fertile land and so on.
3. The WABGs may be invisible for population.
4. Offered installations produce direct electricity. That may be advantage.
5. Many WABGs may be installed along river, falls or sea stream and give big energy.

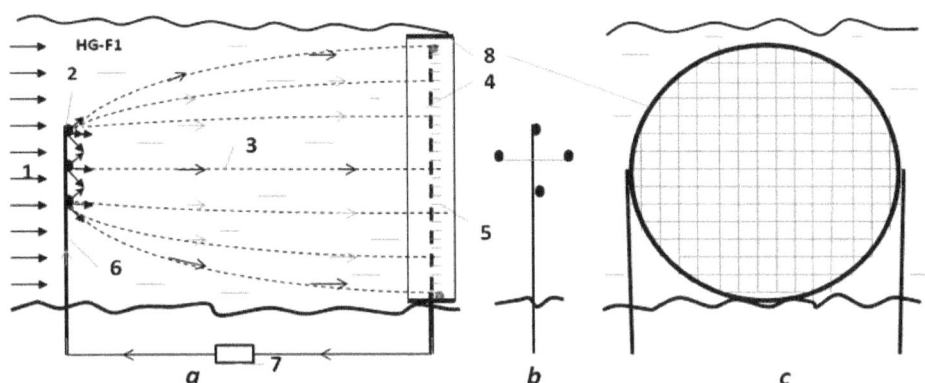

Fig.1. One version of Electron Water Electric Generator (WABG). *a* – side view of the installation; *b* – front view of the electron injector column; *c* – front view of the collect net. *Notations:* 1 is water stream; 2 is electron injector; 3 is trajectories of electrons; 4 is net for collecting the electrons; 5 is conductive leaves (metallic foil, for example, aluminum foil); 6 is column (post) for supporting of the electron injectors; 7 is the outer electric load; 8 – electron high voltage ring collector.

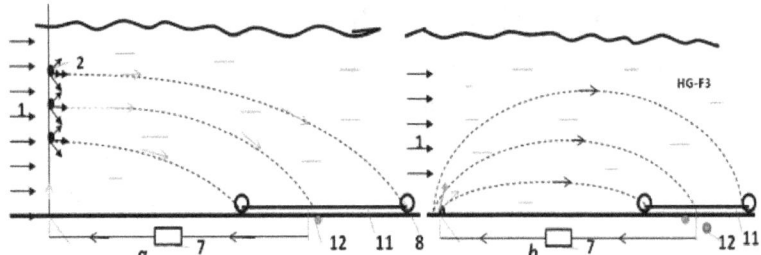

Fig.2. The horizontal conductivity film as collector of electrons. *a* – injectors in column; *b* - injectors at a bottom of river. *Notations:* 1 – 8 are same with fig.1; 11 - conductivity film (for example, aluminum foil); 12 (optional) positive isolated charge (for example, electrets).

Estimations and Computation

1. **Power of a water flow** is N [Watt, Joule/sec]:
$$N = 0.5 \eta \rho A V^3 \quad \text{or} \quad N = \eta \rho B g h \quad [W]. \tag{1}$$

The coefficient of efficiency, η, equals about $0.3 \div 0.5$ for WABG; A - front (forward) area of the electron collector, [m²]. ρ - density of liquid: $\rho \approx 1.000$ kg/m³ for water; V is average annually stream speed, m/s; B is the flow in cubic meters per second; $g = 9.81$ m/s² is Earth gravity; h is the height difference between inlet and outlet of installation.

Example, if $V = 1$ m/s, $A = 1$ m², $\eta = 0.5$, $\rho = 1000$ kg/m³, than $N = 250$ W/m².

The h and V connected by equation
$$h = V^2/2g. \tag{2}$$

Example, if $V = 1$ m/s, than $h = 0.05$ m.

The flow speed of river significantly depends upon width, depth, discharge and elevation. The speed conventionally increases in narrow riverbed and into depth. Speed may be from 0.1 m/s up 3 m/s and more. For example, the Volga has (after dam about Volgograd) speed 1 – 1.5 m/s, width 4 - 7 km and depth 5 - 15 m. The Gulf Stream (in ocean!) has maximal speed 2.5 m/s.

The energy, E, produced in one year is (1 year ≈ 30.2×10^6 work sec) [J]
$$E = 3600 \times 24 \times 350 N \approx 30 \times 10^6 N, \quad [J]. \tag{3}$$

2. Electron speed. The electron speed about the water, wind, gas (air) jet may be computed by equation:
$$j_s = en.b.E + eD.(dn./dx), \tag{4}$$

where j_s is density of electric currency of jet, A/m^2; $e = 1.6 \times 10^{-19}$ C is charge of single electron, C; $n.$ is density of injected electrons (negative charges) in 1 m^3; b . is charge mobility of negative charges, m^2/sV; E is electric intensity, V/m; $D.$ is diffusion coefficient of charges; $dn./dx$ is gradient of charges. For our estimation we put $dn./dx = 0$. In this case
$$j_s = en.b.E, \quad Q_1 = en, \quad v = bE, \quad j_s = Q_1 v, \tag{5}$$

where Q_1 is density of the negative charge in 1 m^3; v is speed of the negative charges about stream, m/s.

One liter of sea water has 35 grams of calt NaCl. The Cl (Chlorine) is 1.9%, the Na (Sodium) is 1.05% of water mass. The calt (saline) dissociates in ions Na$^+$, Cl$^-$. Concentration of ions: Cl$^-$ is 0.546 mol/kg, Na$^+$ is 0.469 mol/kg.

The charge mobility is:
$$\text{Cl}^- \text{ is } 0.667 \times 10^{-7} \text{ m}^2/\text{sV}, \quad \text{Na}^+ \text{ is } 0.450 \times 10^{-7} \text{ m}^2/\text{sV}. \tag{6}$$

As you see the mobility of ions in water is very small. The applied voltage in water is also small. That means the ion speed is small in the comparison with water speed. In many case we can put $v = 0$

If $v > 0$, the electrons accelerate the water ($E > 0$ and installation spends energy, works as engine). If $v < 0$, the electrons brake the water ($E < 0$ and the correct installation can produce energy, works as electric generator). If $v = 0$ (electron speed about installation equals water speed V), the electric resistance is zero.

3. Resistance of water. Salt water conducts an electric currency. That means the part of currency will flows back to cathode. The specific electric resistance of water is significantly depends from salinity of water. When we have the plates (nets) with both sides (cathode and anode), the specific electric resistance are:

1. Distilled water $R \approx 10^6$ Ωm.
2. Fresh water $R = 40 \div 200$ Ωm (depends from water salinity). (7)
3. Sea water $R \approx 0.2$ Ωm.

In our case in one side we have the electron injector (cathode) which has conventionally a small area. In this case the specific electric resistance is:
$$R_o = R / 4\pi a, \tag{8}$$
where a is radius of needle (or cathode), m; this radius conventionally is very small (mm). That means the R_o has an electric resistance of hundreds Ohms. We can neglect their influence in the installation efficiency.

4. The efficiency of installation from back electric currency may be estimated by equation:
$$\eta \approx 1/(1+ R_u/R_o) , \qquad (9)$$
where R_u is an useful electric resistance. Ratio R_u/R_o conventionally is small and η is closed to 1.

5. Specific power of Installation N_1 [W/m²]. The specific power of the offered installation may be estimated by a series of equations:
$$N_1 \approx \eta A_1/t = \eta Q_1 EL/t = \eta Q_1 EV = j_s U = \eta \rho B_1 gh = 0.5\eta m_1 V^2, \qquad (10)$$
where A_1 is energy of flow through 1 m², J/m²; t is time, sec; B_1 is flow in m³ through cross section area of flow 1 m²; E is electric intensity, V/m; L is distance between injector and net (cathode and anode); V is flow speed, m/s; j_s is density of electric currency, A/m²; U is electric voltage, V; m_1 is flow mass per second through area 1 m²; Q_1 is density of the negative charge in 1 m³; g = 9.81 m/s² is Earth gravity; h is the height difference between inlet and outlet of installation (between electron injector and net, between cathode and anode), m.

Projects
Project 1. Small river

Assume we have a small river having the width 100 m, the depth 10 m and speed 1 m/s (slope 0.05), our electronic installation has the electric efficiency η = 0.5 , L = 1 m. That the power of flow through the cross section area 1 m² of flow is
$$N_1 = N/A = 0.5\eta\rho V^3 = 0.5 \cdot 0.5 \cdot 1000 \cdot 1^3 = 250 \text{ W/m}^2 .$$
The total power is
$$N = N_1 A = 250 \cdot 100 \cdot 10 = 250 \text{ kW}.$$
If we install the voltage U = 100 V, the density of electric currency will be
$$j_s = N_1/U = 250/100 = 2.5 \text{ A/m}^2 \quad \text{or} \quad I = j_s A = 2500 \text{ A}.$$
If L = 1 m, the electric intensity is E = U/L = 100 V/m and the difference of water levels is h = 0.05 m. If we take the distance between cathode and anode L = 10 m that for the given electric intensity we can increase the voltage up U = 1000 V, the power of installation up
$$N = 2500 \text{ kW}.$$
But the difference of water levels increases up h = 0.5 m. That is equivalent the small dam h = 0.5 m.

For increasing the power we can increase the distance between cathode and anode (if the size of the river allows it) or use a series of installation along the river.

If we use the simple bottom electron collector (fig.2b), they do not interfere with ships navigation but their electric efficiency may be less.

For Gulf Stream having V = 2.5 m/s the result may be better by a factor of hundreds of times.

Project 2. Niagara Falls

If we install the electron injectors (cathodes) (charged positive) at top level of Niagara Fall and metal sheets and collection ring at bottom (charged negative) we get the electric power up 275 MW (for electric efficiency η = 0.3). Tourists will not see any changes in view of Niagara Fall.

Conclusion

Relatively no significant progress has been made in renewable energy technology in the last years. While the energy from hydro-electric station is free, its building is very expensive (dam, hydro and electric generators) and take a big time. Conventional hydroelectric stations have approached their maximum energy extraction potential relative to their installation cost. They produce the problems for ship navigation, for fish productivity and flood a large area of fertile land. Current large dam installations cannot significantly decrease a cost of kWh.

The hydro-electric energy industry needs revolutionary ideas that improve performance parameters (installation cost) and that significantly decrease (by many times) the cost of energy production. The electron water installations delineated in this paper can move the water energy industry from stagnation to revolutionary potential.

The following is a list of benefits provided by the proposed new electron water systems compared to current dam installations:

1. Offered installations are very simple.
2. Offered system is very cheap (by tens or hundreds of times). No dam, hydro-turbines, electric generators, special canals for ships, fish, filling the fertile land and so on.
3. Many WABGs may be installed along river or sea stream and give big energy.
4. No problems with ships, fish and fertile riversides.
5. The WABGs may be invisible for population.
6. Offered installations produce direct electricity. That may be advantage.

The same method may be applied to tidal, wave and fall powers. As with any new idea, the suggested concept is in need of research and development. The theoretical problems do not require fundamental breakthroughs. It is necessary to design small, cheap installations to study and get an experience in the design electron water generator.

This paper has suggested some design solutions from patent application [2]. The author does not show some important details of this method. He has more detailed analysis in addition to these presented projects. Organizations or investors are interested in these projects can address the author (http://Bolonkin.narod.ru , aBolonkin@juno.com , abolonkin@gmail.com).

The other ideas are in [5]-[7].

References

(Reader can find part of these articles in WEBs: http://Bolonkin.narod.ru/p65.htm, http://www.scribd.com(23); http://arxiv.org , (45), http://vixra.org (15); http://www.archive.org/ (20) and http://aiaa.org (41) and http://intellectualarchive.com, search "Bolonkin").

37. Bolonkin A. A., Utilization of Wind Energy at High Altitude, AIAA-2004-5756, AIAA-2004-5705. International Energy Conversion Engineering Conference at Providence, RI, USA, Aug.16-19, 2004.
38. Bolonkin, A. A, "Method of Utilization a Flow Energy and Power Installation for It", USA patent application 09/946,497 of 09/06/2001.
39. Bolonkin, A. A., Flight Wind Turbines. http://www.scribd.com/doc/138350864/, http://viXra.org/abs/1304.0159
40. Bolonkin A. A., Electronic Wind Generator. http://viXra.org/abs/1306.0046, http://www.scribd.com/doc/146177073/
41. Bolonkin, A. A., "New Concepts, Ideas, Innovations in Aerospace, Technology and the Human Sciences", NOVA, 2006, 510 pgs. http://www.scribd.com/doc/24057071, http://www.archive.org/details/NewConceptsIfeasAndInnovationsInAerospaceTechnologyAndHumanSciences.

42. Bolonkin, A. A., "New Technologies and Revolutionary Projects", Lulu, 2008, 324 pgs, http://www.scribd.com/doc/32744477 , http://www.archive.org/details/NewTechnologiesAndRevolutionaryProjects,
43. Bolonkin, A. A., Cathcart R. B., "Macro-Projects: Environments and Technologies", NOVA, 2007, 536 pgs. http://www.scribd.com/doc/24057930 . http://www.archive.org/details/Macro-projectsEnvironmentsAndTechnologies .
44. Micro-hydro power, Adam Harvey, 2004, Intermediate Technology Development Group, retrieved 1 January 2005.
45. Microhydropower Systems, US Department of Energy, Energy Efficiency and Renewable Energy, 2005.
46. "Hydroelectric Power". Water Encyclopedia. http://www.waterencyclopedia.com/Ge-Hy/Hydroelectric-Power.html
47. Wikipedia. Water Energy.

5 June 2013

Chapter 10

Hydro Propulsion for High Speed Submarines

Abstract.

High speed submarines and in particular **torpedoes** need new propulsion systems which allow the submarine to reach high speeds by cheaper and more efficient methods. Author offers a new propulsion system using electrons for acceleration of the water and having a high efficiency. As this system does not use a water propeller, it does not have the cavitation limitations of conventional water propeller systems. Offered engine can produce a thrust from a zero speed up to high speed. It can work in any liquid planet atmosphere. The system can use apparatus surface for thrust and braking. For energy the system uses high voltage electricity which is not a problem if you have an appropriate electrostatic generator connected with any suitable engine.

Key words: Electron propulsion, WABP, super speed hydro propulsion.

1. INTRODUCTION

Currently, propeller propulsion systems are widely used in submarines. Although they are good for slow speeds (<90 km/h), they are worse for middle speed (> 100 km/h) and has tremendous difficulties in achieving very high speed (> 150 km/h). The current designs of super speed hydro propulsion are limited by cavitations, noise and damage of propeller material.

A **jet engine** is a reaction engine that discharges a fast moving jet which generates thrust by *jet propulsion* in accordance with Newton's laws of motion. They are widely used in aviation, but not a known method in submarines.

2. INNOVATIONS

One simple version of the offered electronic ramjet hydro propulsion (WABP) is shown in fig.1. Engine contains the tube. The ejectors of electrons 2 are installed in the entrance of the tube. The collector of electrons (grille) 3 is installed in the end of tube. The electric circle having the battery (or electric DC generator) 4 and regulator of voltage 7 connects the ejectors and grille.

The engine works the following way. The ejectors eject the electrons into the tube. The strong electric field between injectors and grill moves them to grill. Electrons push (accelerate) the air to tube exit. When the electrons reach the grill, they enter the grill and close the electric circuit. The accelerated water (water jet) with high speed flows out from engine and creates the thrust. In a correctly designed engine this thrust may be enough for moving the craft.

The proposed idea of a propulsion engine has many variations. One of them is shown in fig. 2a. That is a conventional submarine (torpedo) body or wing (in fig. 2a it is shown the gross section of the wing). The electron injectors are installed in the beginning of the body (wing) surface. The collectors are installed in the end of the body/wing. The electrons accelerate the water around the apparatus and the electric forces produce the thrust.

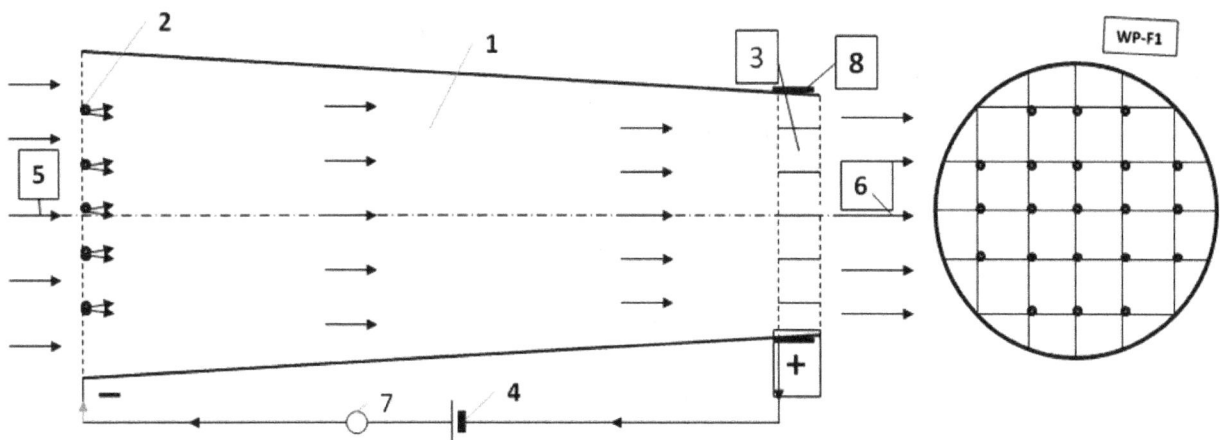

Fig.1. Electron ramjet engine (WABP). *a* – side view, *b* – forward view. *Notations:* 1 – engine; 2 – injectors of electrons; 3 – collector of electrons; 4 – electric issue; 5 – enter water; 6 – exit water jet; 7 – regulator of an electric voltage (electron regulator); 8 – electron collector.

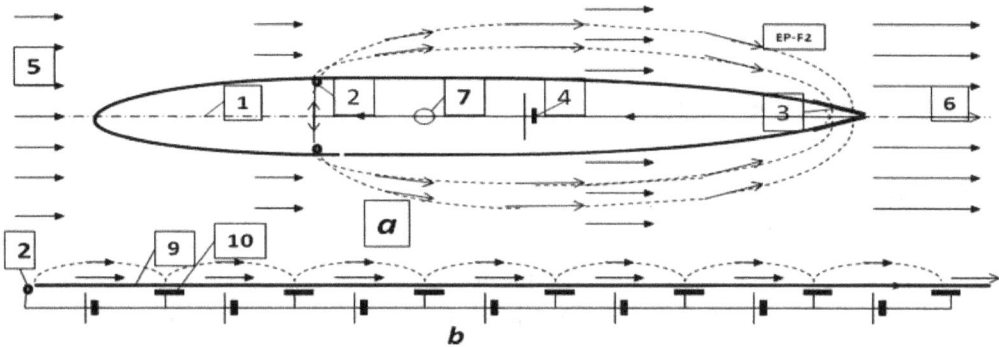

Fig.2. Outer Electron ramjet propulsion (WABP). *a* – side view of the body or a gross-section of wing, *b* – surface electron engine. *Notations:* 1 – fuselage or wing; 2 – injector of electrons; 3 – collector of electrons; 4 – electric issue; 5 – enter water; 6 – exit water jet; 7 – electric (electron) regulator; 9 – surface (isolator) of apparatus; 10 – electric plate.

One possible electric schema of the proposed engine, shown in fig. 3, has additional closed loop electric circles which allow extracting the electrons from main electric circle and collecting electrons from water flow to back into main circle.

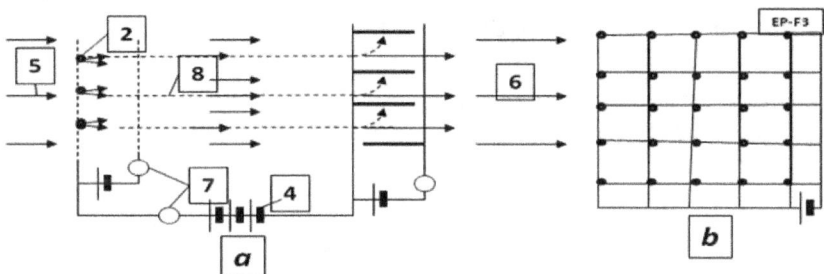

Fig.3. The electrical circuit of one version WABP engine. Notations are same with figs. 1 – 2. *a* is side view, *b* is forward view.

Advantages of the proposed electron propulsion system in comparison with the conventional propeller propulsion systems.

The suggested new propulsion principle has the following advantages in comparison with conventional propeller propulsion systems used at present time.

Advantages:
1. All current water propeller propulsion has low propeller efficiency (about 30% or low for high speed).
    The electronic propulsion engine (WABP) has the electric efficiency about 100% which makes it 3 more times efficient.
2. The electron engine does not have the cavitations and can work at any speed. That means it may be used as an engine of super speed ships.
3. The electronic engine is very simple and cheap.
4. The outer water ship surface may be used as engine. The ship need not have nacelles (propeller-gondolas). That means a high aerodynamic efficiency of sea apparatus.
5. The outer surface of the electronic engine (fig.2b) may be used for creating the laminar boundary layer.
    That means low (minimal) water friction and very high aerodynamic efficiency of sea apparatus.

3. THEORY OF ELECTRON PROPULSION (WABP). COMPUTATION AND ESTIMATION.

1. Thrust of WABP. The thrust of the jet electron engine is (we use the Law of Impulse):

$$T = m(V_f - V) = m\Delta V, \quad m = \rho SV, \quad T = \rho SV\Delta V, \quad T_s = \rho V\Delta V, \quad (1)$$

where T is thrust, N; m is water mass passed through engine in one second, kg/s; V_f is an exit speed of water (medium), m/s; V is an entry speed of water (medium), (the speed of the apparatus), m/s; ΔV is increasing of water (medium) speed into engine, m/s; ρ is water (medium) density, kg/m³; S is ender area of engine, m²; T_s is specific thrust of engine, N/m².

The energy A_t [J] obtained by sea apparatus from thrust is

$$A_t = TVt, \quad (2)$$

where t is time, sec.

On other hand, energy A_e [J] obtained from electric current is

$$A_e = UIt, \quad (3)$$

where U is voltage between entrance and exit of engine, V; I is electric current, A.

The heat efficiency of the WABP is close to 1, because there is no heating of water into engine (increasing the speed of all water mass is in one direction by electric field).
That way

$$A_t \approx A_e. \quad (4)$$

From (1) – (4) and $I_s = I/S$ we get ($V \neq 0$)

$$T_s = \frac{U}{V}I_s, \quad \Delta V = \frac{UI_s}{\rho V^2}, \quad (5)$$

where I_s is density of electric currency about apparatus, A/m², ΔV is increasing water (medium) speed into engine, m/s.

Example 1. Let us take the $U = 10^3$ V, $I_s = 10^3$ A/m², ship speed $V = 50$ m/s, $\rho = 10^3$ kg/m³. Then $T_s = 4 \times 10^4$ N/m² = 4 tons/m², $\Delta V = 0.8$ m/s.

The same way we can get the required power and getting thrust when the ship speed equals zero:

$$P_s = 0.5m \Delta V^2, \quad m = \rho \Delta V, \quad T_s = P_s/\Delta V, \quad P_s = 0.5\rho\Delta V^3, \quad T_s = 0.5\rho\Delta V^2, \quad (6)$$

where P_s is electric power for 1 m², W/m²; ΔV is increasing water speed into engine, m/s; m is water exemption mass passed through engine in one second, kg/s;

Example 2. Let us take the start power $P_s = 10^5$ W/m². Than the exit speed $\Delta V = 2.71$ m/s, the start thrust is $T_s = 3.69 \cdot 10^4$ N/m² = 3.69 tons/m².

2. Efficiency of Electron WABP engine.

Efficiency η of any jet propulsion is production of two values: propulsion efficiency η_p and engine (propeller) efficiency η_e:

$$\eta = \eta_p \eta_e, \quad \text{where} \quad \eta_p = V/(V + 0.5 \Delta V). \quad (7)$$

The propulsion efficiency for propeller and electronic propulsion are same. They depend only on ΔV. But water efficiency of the water propeller is low, about 30 ÷ 40%. In the cavitation regime the propeller efficiency is significantly lower. For high speed over $V > 50$ m/s the conventional water propeller loses efficiency very quickly. The offered electronic jet engine accelerates water (liquid) by electricity. It has efficiency close to 100% as the only loss of energy is the extraction of the electrons from cathode and ionizations of water molecules. This energy is about some (about 0) electron-volts (eV). The energy spent for acceleration of the water molecules by electrons/ions is hundreds of eV. That means the total efficiency of WABP is 3 times more than conventional air jet propulsion.

The second very important point: efficiency of WABP does not depend upon speed of apparatus.

The other advantages: we can make a very large entrance area of engine, we can use the fuselage and wings, stabilizer and keel of ship as engine.

3. Electron speed.
The electron speed about the water, wind, gas (air) jet may be computed by equation:

$$j_s = en_-b_-E + eD_-(dn_-/dx), \quad (8)$$

where j_s is density of electric currency of jet, A/m²; $e = 1.6 \times 10^{-19}$ C is charge of single electron, C; n_- is density of injected electrons (negative charges) in 1 m³; b_- is charge mobility of negative charges, m²/sV; E is electric intensity, V/m; D_- is diffusion coefficient of charges; dn_-/dx is gradient of charges. For our estimation we put $dn_-/dx = 0$. In this case

$$j_s = en_-b_-E, \quad Q_1 = en_-, \quad v = bE, \quad j_s = Q_1 v, \quad (9)$$

where Q_1 is density of the negative charge in 1 m³; v is speed of the negative charges about stream, m/s.

One liter of sea water has 35 grams of calt NaCl. The Cl (Chlorine) is 1.9%, the Na (Sodium) is 1.05% of water mass. The salt (saline) dissociates in ions Na^+, Cl^- into water. Concentration of ions: Cl^- is 0.546 mol/kg, Na^+ is 0.469 mol/kg.

The charge mobility is:

$$Cl^- \text{ is } 0.667 \times 10^{-7} \text{ m}^2/\text{sV}, \quad Na^+ \text{ is } 0.450 \times 10^{-7} \text{ m}^2/\text{sV}. \quad (10)$$

As you see the mobility of ions in water is very small. The applied voltage in water is also small. That means the ion speed is small in comparison with water speed. In many case we can put $v = 0$.

If $v > 0$, the electrons accelerate the water ($E > 0$ and installation expends energy, works as engine). If $v < 0$, the electrons brake the water ($E < 0$ and in the correct installation can produce energy, works as electric generator). If $v = 0$ (electron speed about installation equals water speed V), the electric resistance is zero.

4. Resistance of water. Salt water conducts electric current. This means that part of current will flow back to cathode. The specific electric resistance of water is significantly contingent upon salinity of water. When we have the plates (nets) with both sides (cathode and anode), the specific electric resistance are:

1. Distilled water $R \approx 10^6$ Ωm.
2. Fresh water $R = 40 \div 200$ Ωm (depends from water salinity). (11)
3. Sea water $R \approx 0.2$ Ωm.

In our case in one side we have the electron injector (cathode) which has conventionally a small area. In this case the specific electric resistance is:

$$R_o = R/4\pi a, \qquad (12)$$

where a is radius of needle (or cathode), m; this radius conventionally is very small (mm). That means the R_o has an electric resistance of hundreds Ohms. Their influence in the installation efficiency is insignificant.

4. The efficiency of installation from back electric current may be estimated by equation:

$$\eta \approx 1/(1+ R_u/R_o), \qquad (13)$$

where R_u is an useful electric resistance. Ratio R_u/R_o conventionally is small and η is closed to 1.

4. SUMMARY AND DISCUSSION.

The author proposes a fundamentally unique propulsion system (engine) using the outer medium (water, liquid) and electric energy. It is not comparable to conventional propeller propulsion which uses the mechanical engine for rotation of propeller.

The offered WABP engine is accelerating the water (medium) by a principally new method – by electric field. Its advantages:
1. All current water propeller propulsion have low propeller efficiency (about 30% or low for high speed).
   The electronic propulsion engine (WABP) has electric efficiency about 100% which makes it 3 more times efficient.
2. The electron engine does not have cavitations and thus can work at any speed. This means that it may
   be used as an engine of super speed ships.
3. The electronic engine is very simple and cheap.
4. The outer water ship surface may be used as engine. The ship need not have nacelles (propeller-gondolas). That means high aerodynamic efficiency of the sea apparatus.
5. The outer surface electronic engine (fig.2b) may be used for creating the laminar boundary layer. That means low (minimal) water friction and very high aerodynamic efficiency of sea apparatus.

This engine is also dissimilar to a known propeller or rocket engines (underwater rocket used by super speed torpedo). The rocket engine takes an extracted mass from itself. The WABP does not take the extracted mass, can work only in water and works better if the atmosphere has a high density.

The main disadvantage of the offered engine is the requirement of electricity. For obtaining electricity the conventional internal turbo engine connected with electro-DC generator may be used.

The research papers relating to this topic are presented in [1]-[20].

References

[1] A.A.Bolonkin, Electron Air Hypersonic Propulsion. http://viXra.org/abs/1306.0003
[2] A.A.Bolonkin, Electronic Wind Generator. http://viXra.org/abs/1306.0046,
[3] A.A.Bolonkin, Electronic Hydro Electro Generator. 2013.
[4] A.A. Bolonkin, "Bolonkin's Method Movement of Vehicles and Installation for It", US Patent 6,494,143 B1, Priority is on 28 June 2001.
[5] A.A. Bolonkin, "Non-Rocket Space Launch and Flight", Elsevier, 2005, 468 pgs. ISBN-13: 978-0-08044-731-5, ISBN-10: 0-080-44731-7 . http://www.archive.org/details/Non-rocketSpaceLaunchAndFlight , http://www.scribd.com/doc/24056182
[6]. A.A. Bolonkin, "*New Concepts, Ideas, Innovations in Aerospace, Technology and the Human Sciences*", NOVA, 2006, 510 pgs. ISBN-13: 978-1-60021-787-6. http://www.scribd.com/doc/24057071 , http://www.archive.org/details/NewConceptsIfeasAndInnovationsInAerospaceTechnologyAndHumanSciences
[7]. A.A. Bolonkin, R. Cathcart, "*Macro-Projects: Environments and Technologies*", NOVA, 2007, 536 pgs. ISBN 978-1-60456-998-8. http://www.scribd.com/doc/24057930 . http://www.archive.org/details/Macro-projectsEnvironmentsAndTechnologies .
[8]. A.A. Bolonkin, *Femtotechnologies and Revolutionary Projects*. Lambert, USA, 2011. 538 p., 16 Mb. ISBN: 978-3-8473-0839-0. http://www.scribd.com/doc/75519828/ , http://www.archive.org/details/FemtotechnologiesAndRevolutionaryProjects
[9]. A.A. Bolonkin, *LIFE. SCIENCE. FUTURE* (Biography notes, researches and innovations). Scribd, 2010, 208 pgs. 16 Mb. ISBN: 978-1-4512-7983-2, http://www.scribd.com/doc/48229884, http://www.archive.org/details/Life.Science.Future.biographyNotesResearchesAndInnovations
[10]. A.A. Bolonkin, *Universe, Human Immortality and Future Human Evaluation*. Scribd. 2010г., 4.8 Mb. http://www.archive.org/details/UniverseHumanImmortalityAndFutureHumanEvaluation, http://www.scribd.com/doc/52969933/
[12]. A.A.Bolonkin, "Magnetic Space Launcher" has been published online 15 December 2010, in the ASCE, *Journal of Aerospace Engineering* (Vol.24, No.1, 2011, pp.124-134). http://www.scribd.com/doc/24051286/
[13]. A.A.Bolonkin, Universe (Part 1). Relations between Time, Matter, Volume, Distance, and Energy. JOURNAL OF ENERGY STORAGE AND CONVERSION, JESC : JuLy-December 2012, Volume 3, Number 2, pp. 141-154. http://viXra.org/abs/1207.0075, http://www.scribd.com/doc/100541327/ , http://archive.org/details/Universe.RelationsBetweenTimeMatterVolumeDistanceAndEnergy
Universe (Part 2): Rolling of Space (Volume, Distance), Time, and Matter into a Point. http://www.scribd.com/doc/120693979
http://viXra.org/abs/1207.0075, http://www.scribd.com/doc/100541327/ , http://archive.org/details/Universe.RelationsBetweenTimeMatterVolumeDistanceAndEnergy
[14]. A.A.Bolonkin, Lower Current and Plasma Magnetic Railguns. Internet, 2008. http://www.scribd.com/doc/31090728 ; http://Bolonkin.narod.ru/p65.htm .
[15] A.A.Bolonkin, Electrostatic Climber for Space Elevator and Launcher. Paper AIAA-2007-5838 for *43 Joint Propulsion Conference*. Chincinnati, Ohio, USA, 9 – 11 July,2007. See also [10], Ch.4, pp. 65-82.
[16] W.J. Hesse and el. Jet Propulsion for Aerospace Application, Second Edition, Pitman Publishing Corp. NY.
[17] N.I. Koshkin and M.G. Shirkebich, Directory of Elementary Physics, Nauka, Moscow, 1982 (in Russian).
[18] I.K. Kikoin. Table of Physics values. Atomisdat, Moscow, 1976 (in Russian).
[19] Water Encyclopedia. http://www.waterencyclopedia.com/Ge-Hy/Hydroelectric-Power.html
[20] Wikipedia. Submarines.

23 June 2013

Short biography of Bolonkin, Alexander Alexandrovich

Alexander A. Bolonkin was born in the former USSR. He holds doctoral degree in aviation engineering from Moscow Aviation Institute and a post-doctoral degree in aerospace engineering from Leningrad Polytechnic University. He has held the positions of senior engineer in the Antonov Aircraft Design Company and Chairman of the Reliability Department in the Clushko Rocket Design Company. He has also lectured at the Moscow Aviation Universities. Following his arrival in the United States in 1988, he lectured at the New Jersey Institute of Technology and worked as a Senior Scientist at NASA and the US Air Force Research Laboratories.

Bolonkin is the author of more than 250 scientific articles and books and has 17 inventions to his credit. His most notable books include The Development of Soviet Rocket Engines (Delphic Ass., Inc., Washington , 1991); Non-Rocket Space Launch and Flight (Elsevier, 2006); New Concepts, Ideas, Innovation in Aerospace, Technology and Human Life (USA, NOVA, 2007); Macro-Projects: Environment and Technology (NOVA, 2008); Human Immortality and Electronic Civilization, 3-rd Edition, (Lulu, 2007; Publish America, 2010):Life and Science. Lambert Academic Publishing, Germany, 2011, 205 pgs. ISBN: 978-3-8473-0839-3. http://www.archive.org/details/Life.Science.Future.biographyNotesResearchesAndInnovations; Femto technology and Revolutionary {rojectsts, Lambert, 2011, p.530; New Methods of Optimization and their Application, Moscow High Technical University named Bauman (in Russian: Новые методы оптимизации иих применение. МВТУим. Баумана, 1972г., 220 стр). List and links of Bolonkin's publication: http://viXra.org/abs/1604.0304.Homepage: http://Bolonkin.narod.ru.

In given book considered the topics: utilization wind energy at high altitude, transwer of energy from airborne wind turbines to ground surface, new non turbine electron wind and water electric generators and propulsion system.

Author offers a new method of getting electric energy from wind. A special injector injects electrons into the atmosphere. Wind picks up the electrons and moves them in the direction of wind which is also against the direction of electric field. At some distance from injector a unique grid acquires the electrons, thus charging and producing electricity. This method does not require, as does other wind energy devices, strong columns, wind turbines, or electric generators. This proposed wind installation is cheap. The area of wind braking may be large and produces a great deal of energy. Although this electron wind installations may be in a city, the population will not see them.

Author offers a new high efficiency propulsion non turbine system using electrons for acceleration of the craft. As this system does not heat the air, it does not have the heating limitations of conventional air ramjet hypersonic engines. Offered engine can produce a thrust from a zero flight speed up to the desired escape velocity for space launch. It can work in any planet atmosphere (gas, liquid) and at high altitude.

www.ingramcontent.com/pod-product-compliance
Lightning Source LLC
Chambersburg PA
CBHW080917170526
45158CB00008B/2148